Ethically Speaking

Essays and Other Writings on Race, Class and Justice in Health Care and Medicine

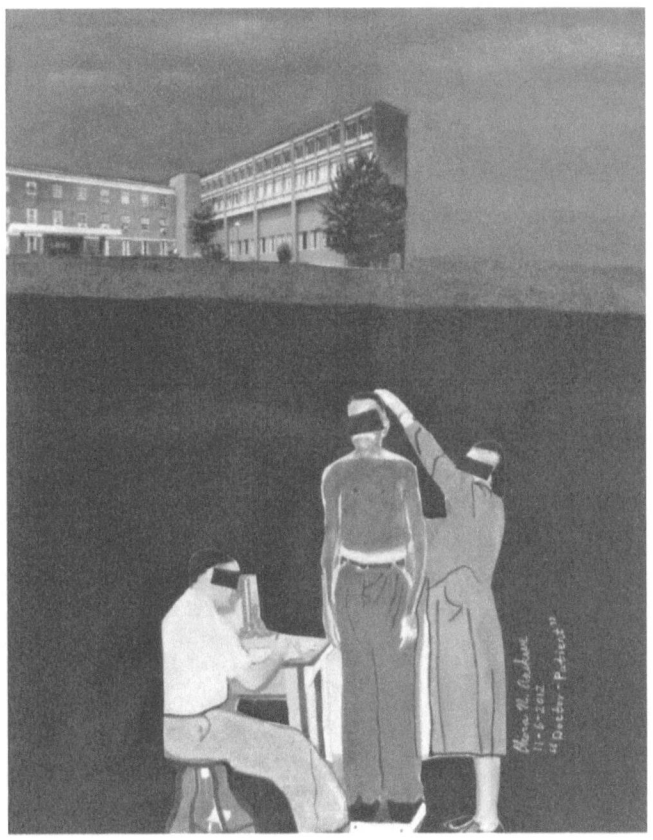

Obiora N. Anekwe, Ed.D, M.S (Bioethics)

Ethically Speaking

Essays and Other Writings on Race, Class
and Justice in Health Care and Medicine

Obiora N. Anekwe, Ed.D, M.S (Bioethics)

Copyright © 2014 Obiora N. Anekwe, Ed.D, M.S (Bioethics)
All rights reserved.

ISBN: 1497531217
ISBN 13: 9781497531215
Library of Congress Control Number: 2014906496
CreateSpace Independent Publishing
North Charleston, South Carolina

About the Cover Art

The front cover art collage entitled, "Doctor-Patient," shows a doctor and Nurse Eunice Rivers weighing and confirming the height of a male participant in the Tuskegee Syphilis Study. The upper left corner of the collage shows the John A. Andrew Memorial Hospital on the campus of Tuskegee Institute (now Tuskegee University), where the study was conducted. The faces of the doctor, nurse, and study participant are obscured to emphasize that their personal identities are not as significant as their participation in the study. No matter the role of the individual, each person involved in the study—doctor, nurse, participant—is in some way ignorant of the consequences of his or her actions.

During the study, which lasted from 1932 to 1972, doctors from the US Public Health Service studied the progression of untreated syphilis in rural African American men who were misled into believing they were receiving free health care. Doctors involved in the study abused their role as medical experts to direct participants, updating records with their weight and physical condition and regularly testing their blood to check the syphilitic status of participants who were in the nonsyphilitic group. Yet, throughout the forty-year span of the study, the participants never provided consent or received any explanation for why they had to endure painful tests, such as spinal taps done without anesthesia. These unethical medical actions eventually caused physical harm and even death for some participants.

The doctors who conducted the Tuskegee Syphilis Study defended their deceptive actions as necessary for the greater good of society. For example, although penicillin was a known treatment for syphilis in the 1940s, doctors withheld this treatment because they wanted to observe the effects of nontreatment of syphilis in the African American male body. Their justification for nontreatment was based on a belief that if the nontreatment could be observed, then medical doctors could prove that the black male body reacted to a sexually transmitted disease differently than the white male body.

I was inspired to create this art collage because I was born in the John A. Andrew Memorial Hospital. As a young child, I was somewhat aware of the study because my parents worked at Tuskegee Institute during the 1960s when the clinical trial was still being conducted. When I grew older, I wanted to learn more about the historical context of the study. The African American male participants were intentionally deceived about their medical treatment and uninformed about the true intentions of the Tuskegee Syphilis Study. Why did this happen, and how could it have continued for so long? As a bioethicist, I was driven to pursue these questions through my educational and research endeavors. This pursuit has inspired and enriched my life both professionally and personally.

Obiora N. Anekwe, Ed.D, M.S (Bioethics)
November 6, 2012
Brooklyn, New York

> 16 by 20 inches
> Canvas paper, gouache paint, soft pastels, black and white charcoal pencils, colored markers, and copy paper
>
> Original images provided by the Tuskegee University Archives, Tuskegee, Alabama, and the National Archives and Records Administration, Southeast Region, Morrow, Georgia.

About the Back Cover Art

The back cover art collage entitled, "Treatment Seeker," chronicles how before the inception of the Tuskegee Syphilis Study, many women in the Macon County and surrounding areas of Tuskegee, Alabama, sought treatment for syphilis. The local community often referred to treatment for syphilis as a form of treatment for "bad blood." Therefore, many African American women who feared that they might have syphilis or "bad blood" sought treatment by providing blood samples to health-care practitioners who later tested these samples in order to determine if they had syphilis. Although these women who provided blood samples to medical doctors were not treated for syphilis, they were medically deceived to believe that they were being treated for the disease.

Once the official beginning of the Tuskegee Syphilis Study occurred in 1932, African American women were excluded from participating in this clinical trial. Only African American men from the region were enrolled in the Tuskegee Syphilis Study. Even though these men were not treated for syphilis in the study, they continued to participate in the study because they were medically deceived to believe that they were actually being treated for the disease when, in fact, they were not. As a result, many of these men unknowingly infected their wives and loved ones during their tenure as participants in the study. Thus, syphilis also became a generational and deadly disease that consequentially affected multiple families throughout many generations.

The collage shows an African American woman who provided a blood sample to an African American physician in order to test the sample to determine if she had syphilis. Very little research and discussion has focused on the extent to which African American health officials and administrators played a role in perpetuating the Tuskegee Syphilis Study. The woman believed that if she had syphilis, she would be treated for the disease. But unfortunately, like in many similar cases, she was not. This collage serves as a visual reminder of the medical, historical, and sociological impact of nontreatment for syphilis that has consciously and subconsciously affected the unwillingness of many African Americans to participate in public health clinical trials.

A lack of formal education and health-care training contributed to the promotion of nontreatment for many victims of the Tuskegee Syphilis Study. Health-care education, in particular, could have played a vital role in preventing many African American men from participating in this unethical study. The photograph of the Shiloh-Rosenwald School in Notasulga, Alabama, highlights the importance of education and the role it plays in promoting healthy, rational, and informed decision making. Ironically, many children who attended the Shiloh-Rosenwald School would often see family members stationed near the school providing blood samples to public health doctors in order to test for syphilis. The formalized education of schoolchildren in a schoolhouse took place in the near vicinity of the formalized medical deceit of many local children's parents due to a lack of education.

Obiora N. Anekwe, Ed.D, M.S (Bioethics)
May 16, 2014
Brooklyn, New York

> 16 by 20 inches
> Canvas paper, colored paper, white charcoal pencil, black charcoal pencil, black pencil, earth-tone pastels, oil paint, burlap and mixed media cloth material, colored pencils, and gouache paint
>
> Original image of school photographed by the artist.

Additional image provided by the National Archives and Records Administration, Southeast Region, Morrow, Georgia.

Additional back cover image extracted from the collage, "Tuskegee Men," by Obiora N. Anekwe.

Dedicated in loving memory to my grandparents,
SFC (Ret.) Arthur A. Maddox and Mrs. Eliza W. Maddox

Healing is a matter of time, but it is sometimes also a matter of opportunity.

—Hippocrates

About the Author

Photographer: Grace Bello

Obiora N. Anekwe was born in Alabama on the campus of the Tuskegee Institute (now Tuskegee University) in the John A. Andrew Memorial Hospital. Reared in Lagos, Nigeria, he attended the University of Lagos Staff School as a child. He is a graduate of Clark Atlanta University (B.A, mass media arts with honors), Tuskegee University (M.Ed, counseling and student development), Auburn University (Ed.D, educational leadership), and Columbia University in the City of New York (M.S, bioethics with high honors and distinction). His early educational experiences influenced his commitment to learning, presenting, and conducting research in international education in Germany, Italy, Poland, and Scotland.

Before moving to New York, Obiora previously worked as a counselor, instructor, and educational coordinator in the office of the provost at Tuskegee University in Tuskegee, Alabama. In addition, he has taught undergraduate teacher education students as an adjunct instructor in educational foundations at Auburn University in Auburn, Alabama. Obiora has received additional training in bioethics, health care, and clinical ethics from the Kennedy Institute of Ethics at Georgetown University, Union Graduate College, and Mount Sinai School of Medicine. He is a member of the American Society for Bioethics and Humanities, the New York Academy of Sciences, and the American Chemical Society.

Obiora has published in the fields of education and bioethics. His first book, *Celebrating Life at 24 Hampton Place*, documented the

Igbo traditional life and home-going celebrations of his great aunt, Catherine N. Anekwe. His second book, *Chronicling the Tuskegee Syphilis Study: Essays, Research Writings, Commentaries, and Other Documented Works* was cowritten with his brother, Ejinkonye C. Anekwe, Ph.D. Obiora's third book, *Ancestral Voices Rising Up: A Collage Series on the Tuskegee Syphilis Study*, was published in 2014. As a founding contributing editor and writer for the Columbia University online bioethics journal, *Voices in Bioethics*, Obiora has written art reviews on public health and bioethics-related issues, in addition to his writings on race, gender, and vulnerable populations in human subjects research. In addition to his duties as contributing editor and writer, Anekwe also serves as coordinator for the art and bioethics section of *Voices in Bioethics*.

Dr. Obiora N. Anekwe resides in Brooklyn, New York with his wife, Rev. Alexis Southerland Anekwe.

CONTENTS

Preface xix

Chapter 1: Issues of Race, Justice, Global Colorism, Skin Bleaching, and Internalized Perceptions of Skin Color 1

    Race and Justice Ethically Examined
through a Systematic Review of the
Tuskegee Syphilis Study 3

    Global Colorism: An Ethical Issue
and Challenge in Bioethics 77

    The Global Phenomenon of Skin Bleaching:
A Crisis in Public Health (Part I) 98

    The Miseducation of Global Perceptions
of the Negative Effects of Skin Bleaching (Part II) 104

    The Beauty of Lupita Nyong'o and
Internalized Perceptions of Skin Color 110

    We Are Our Brother's Keeper 112

    Healing Voices of a People: Harry Belafonte
and the Belafonte Folk Singers' Exquisite
Interpretation of the Negro Spirituals and a
Movement toward Modern Justice-Based Ethics 114

Chapter 2: A Television Soap Opera's Critique of Issues Related to Health-Care Inequality and Risky Sexual Behavior Practices 119

    Tyler Perry's *The Haves and the Have Not*s:
Infusing Thematic Discussions of Health Care

and Class Disparities through the
Visual Medium of Television (Part I) 120
Tyler Perry's *The Haves and the Have Nots*: Addressing
Forced Duplicity in Black Male Identity
and Risky Sexual-Behavior Practices (Part II).........................122

**Chapter 3: Health Care and Medical Issues
in Urban Public Health** 125
Rising Up: Hale Woodruff's Murals from
Talladega College Travel the Country 126
Urban Ethics and Cultural Moralism
Translated in Spike Lee's *Do the Right Thing*:
Twenty-Five Years Later..128

PREFACE

Medical and health-care ethics can be universal in both nature and scope. But they are also a result of race, class, and justice challenges. My book focuses on highlighting these various issues and how they, ultimately, can be resolved. Chapter 1 consists of my graduate thesis that I completed in bioethics at Columbia University in May of 2014, articles on global colorism and ethical challenges related to skin bleaching (a two-part series), internalized perceptions of beauty, the concept of universal human responsibility, and the justice-based ethical power of Negro spirituals. In chapter 2, I discuss how the television soap opera, *The Haves and Have Nots*, critiques health care and medical challenges, such as health-care inequality and risky sexual behavior practices. In conclusion, chapter 3 focuses on health-care and medical issues in urban public health.

Except for my graduate thesis, most of my chapter articles were previously published in Columbia University's bioethics journal, *Voices in Bioethics*. All chapters within this book are reflective of my research writings as a graduate student in bioethics at Columbia University (Fall 2012 to Spring 2014) and after my recent graduation in May of 2014. My greatest hope is that every person who reads my book will acquire a much broader perspective of issues critically important to people of color, who have been significantly marginalized and most critically affected by race, justice, and class inequalities.

—Obiora N. Anekwe, Ed.D, M.S (Bioethics)

Chapter 1:
Issues of Race, Justice, Global Colorism, Skin Bleaching, and Internalized Perceptions of Skin Color

RACE AND JUSTICE ETHICALLY EXAMINED THROUGH A SYSTEMATIC REVIEW OF THE TUSKEGEE SYPHILIS STUDY

by
Obiora Nnamdi Anekwe

James Colgrove, PhD, MPH, Advisor

A thesis submitted in partial fulfillment
of the requirements for the
degree of master of science
in bioethics

Columbia University

New York, New York

May 21, 2014

He has told you, O man, what is good; and what does the LORD require of you but to do justice, and to love kindness, and to walk humbly with your God?

—Micah 6:8
Biblical English Standard Version

Wake up every day doing something to help enhance justice!

—Harry Belafonte
Icon/Actor/Justice Rights Activist

The unexamined life is not worth living.

—Socrates
Greek Philosopher

Justice is *better* than *Racism*.

—Neely Fuller, Jr.
Race Theorist

CONTENTS

I. Introduction An Overview 9

II. Race 11
    A. Historical Background, Context, and Origins
    B. Race Matters
        What Does Race Have to Do with It?
    C. Race Classification
        1. Colorism within the Tuskegee Community
        2. The US Bureau of Census and Race Classification Structures

III. Justice 27
    A. Justice as Defined Historically through the *Belmont Report*
    B. Two Philosophical Concepts of Justice according to John Rawls and John Locke
    C. The Nature and Purpose of the Institutional Review Board (IRB) as a Means to Enhance Distributive Justice in Human Experimentation
        1. Informed Consent as a Tool to Enhance Distributive Justice in Human Experimentation
        2. Display 1: Distributive Justice Enhancement Model

IV. Race-Based Justice 36
    A. Conceptual Definition and Discussion: Does Race-Based Justice Exist? And Should Justice Be Race Based?

 B. Black Soul Repair (BSR):
   A Holistic Model for Restorative Justice
 C. Display 2: Black Soul Repair Model

V. Conclusions 42
 A. Ethical Recommendations for Solutions
   to Race Challenges in Biomedical Research
 B. Closing Statement

References 59

Appendices 65
 A. Applied Experiences as an IRB Graduate Intern
 B. Display 3: The Black Church Healing Model
 C. Additional Documents Attached

Introduction
An Overview

    The Tuskegee Syphilis Study has been examined in the fields of bioethics, public health, and medicine as a case study of *how not to conduct* human subject experimentation. The study's relevance has reached beyond the borders of academic disciplines in order to address ethical challenges in biomedical research. Until recently, central issues concerning colorism and justice relevant to the Tuskegee Syphilis Study still have not been adequately examined. Although race, in particular, has been examined by many scholars of the Tuskegee Syphilis Study, a critical analysis of colorism within the black community at Tuskegee has not been explored as an explanation as to why and how the Tuskegee Syphilis Study was conducted. My thesis uniquely contributes to literature on the Tuskegee Syphilis Study because it connects how race and justice play significant roles in framing an accurate portrayal of what happened to the victims of the Tuskegee Syphilis Study. I contend that in order to accurately rectify these past unethical medical practices, a clear understanding of race and justice is necessary.

    My thesis is divided into five sections: an introduction, race, justice, race-based justice, and a conclusion. In the first section of my thesis, I coherently introduce the topic at hand. The second section lays a foundation to discuss matters concerning race as they relate to the Tuskegee Syphilis Study. I plan to argue that race, essentially, matters in how justice is ethically conducted. In order to address the ethical concerns that were prevalent within the Tuskegee Syphilis Study, issues of race must be historically brought to the forefront. Hence, I plan to discuss how colorism and the rise of the mulatto class within Tuskegee's black community played a critical role in the recruitment of mostly black male farmers in the study. Additionally, I discuss the role of the US Bureau of the Census in influencing race classifications within the black community.

In the third section, I first define justice according to the *Belmont Report* and then explore the philosophical concept of justice according to John Rawls and John Locke. I also discuss the nature and purpose of the institutional review board (IRB) as a means to enhance distributive justice in human experimentation. In the conclusion of this section, I provide literature on informed consent as a tool to enhance distributive justice in human experimentation.

The fourth section argues that in order to render justice fairly and coherently, the concept of race-based justice must play a pivotal role in responding to unethical medical practices conducted by clinical researchers in the Tuskegee Syphilis Study. I first define race-based justice and provide a discussion as to how this concept can be applied to the Tuskegee Syphilis Study. In the fifth and last section, I conclude by providing ethical recommendations for solutions to race and justice challenges in bioethics. In the appendices section, I highlight how my applied experiences as an IRB graduate intern in bioethics presented ways in which race and justice should be adequately considered in making ethical decisions affecting minority populations such as those studied in the Tuskegee Syphilis Study.

Race
Historical Background, Context, and Origins

The longest clinical trial in the history of North America took place in Macon County in the southern city of Tuskegee Institute, Alabama, from 1932 to 1972. For the purpose of the paper, I shall refer to this clinical trial as the Tuskegee Syphilis Study or the experiments at Tuskegee. The syphilis study was housed at the Tuskegee Institute (now known as Tuskegee University) in the campus clinical facility of the John A. Andrew Memorial Hospital. It was officially known as the *US Public Health Service Syphilis Study at Tuskegee*. The US Public Health Service staff recruited hundreds of African American men in rural Alabama over a forty-year period in order to study and clinically measure the effects of syphilis in the black male body. "Early in the study, 399 men with late latent syphilis, and 201 men without syphilis were initially enrolled. As the study evolved, additional participants were added, so the number of men in the study varies according to the source" (Centers for Disease Control and Prevention 2014, 1).

In order to test the doctors' scientific hypothesis that the African American male body would react differently to syphilis than the white male body, public health doctors did not treat the men for syphilis, but rather medically deceived them in order to measure the validity of their hypothesis (Edgar 1992). Even after penicillin became widely available in the 1940s, treatment was still withheld from the men in order to measure the effects of syphilis until death. A number of local African American and white physicians were recruited for the study in order to prevent treatment of the men (Centers for Disease Control and Prevention 2014). Additionally, a number of Tuskegee Institute faculty and staff were paid by the Public Health Service in order to formally serve as human subject recruiters, researchers, and medical personnel for the study (2014). Many of the men died from the effects the disease had on their physical bodies.

A brief overview of race-based medicine and the prevailing stereotypes about African Americans is necessary in order to fully grasp the racialized components of research and human experimentation during the advent of the study. Social Darwinism and, later, the American eugenics movement began to take hold in American culture at the turn of the century. Scientists speculated that black people were particularly prone to crime and diseases, especially sexually transmitted ones (Brandt 1978). It was argued that "primitive people," such as blacks, could not be assimilated into the complex structure of white society.

By the nineteenth and early twentieth centuries, biologists supported the belief that freedom and emancipation after slavery caused the mental, moral, and physical deterioration of the black race (Deterioration of the American Negro, 1903). This stereotype was based on comparative anatomy studies often cited by notable scientists and physicians such as W. H. English. The interest in racial differences between blacks and whites expanded in order to focus more on the sexual prowess and nature of black sexuality. As one distinguished physician observed in the *Journal of the American Medical Association*, "The negro springs from a southern race, and as such his sexual appetite is strong; all of his environments stimulate his appetite, and as a general rule his emotional type of religion certainly does not decrease it" (Hazen 1914, 463). In particular, the oversexualized nature of the black male was of interest to those who practiced racialized medicine.

The physical and mental inferiority of blacks could not be empirically proven. Rather, it was based on preconceived beliefs by mostly southern doctors who believed in the established social order (Jones 1993). During slavery, these physicians justified the enslavement of blacks and, even after its extinction, "second-class citizenship by insisting that blacks were incapable of assuming any higher station in life" (1993, 17). In the minds of these physicians, there existed too many physical and mental differences between blacks and whites. These supposed differences, in essence, meant that blacks were medically inferior (1993). This unfounded belief gave rise to the rationale that blacks should be kept in their place for their own good and the good of society (1993).

Doctors also argued and feared the black males' possible desire, craving, and lust for white women. As Dr. English once wrote, "A perversion from which most races are exempt prompts the negro's inclination towards white women, whereas other races incline towards females of their own" (Hazen 1914, 463). He further claimed the "gray matter of the negro brain" to be at least a thousand years behind that of the white race and his genital organs to be overdeveloped (1914, 463).

Therefore, according to physicians of the 1800s and even until the middle 1900s, lust, immorality, and a possible reversion back to barbaric tendencies made black people, especially males, more prone to venereal diseases. Doctors believed that any treatment for venereal disease among black people was virtually impossible because, in its latent stage, the symptoms of diseases, such as syphilis, become quiescent (Brandt 1978). Furthermore, according to medical opinion of the time, blacks were once free of disease as slaves but virtually overwhelmed by it due to their newfound freedom after enslavement (1978). There was a fervent belief among physicians that venereal disease was inherent to the black population. Doctors contended that even the best of medical care could not alter the evolutionary scheme ("As Ye Sow, That Shall Ye also Reap," 1899). They simply discounted the socioeconomic and educational disparities that existed then and still exist today in the African American community. This particular configuration of unfounded medical opinions laid a solid foundation for developing an experiment to measure the effects of nontreatment for syphilis in the black male, known now as the Tuskegee Syphilis Study.

Race Matters: What Does Race Have to Do with It?

One of the least explored means by which the Tuskegee Syphilis Study has been historically framed is through the lens of colorism and classism within science. By all accounts, the Tuskegee Syphilis Study was a *race-based* study with *race-based* implications that affected its participants as well as generations to come. Thus, it is my goal to begin a conversation on the issue of race as identified within the syphilis study. As such, evidence suggests that race is "an arbitrary system of visual classification that does not demarcate distinct subspecies of the human population" (Fullilove 1998). In fact, as Fullilove (1998) explains, the concept of race was "developed largely to justify the highly profitable African slave trade and the systems of slavery in the Americas, hinged on the 'natural inferiority' of 'colored' peoples to 'Whites'" (Fullilove 1998). As evident from the Tuskegee Syphilis Study, economic and class inequalities also create divergent health outcomes for many populations, most notably African American and Hispanic-American populations.

ISSUES OF RACE, JUSTICE, GLOBAL COLORISM, SKIN BLEACHING, AND INTERNALIZED PERCEPTIONS OF SKIN COLOR

But in order to clearly begin a discussion on race as it relates to the Tuskegee Syphilis Study, some socially constructive terms need to be defined. Accordingly, *racism* refers to "institutional and individual practices that create and reinforce oppressive systems of race relations whereby people and institutions engaging in discrimination adversely restrict, by judgment and action, the lives of those against whom they discriminate" (Krieger 2003, 195). In turn, race is "a social rather than a biological category, referring to social groups, often sharing cultural heritage and ancestry, that are forged by oppressive systems of race relations justified by ideology" (2003, 195). Nonetheless, race, once viewed as a scientific fact, is in fact refuted by research in genetics, anthropology, and sociology to be invalid biologically (2003). Henceforth, race is more about our society than biology itself (2003).

Researchers who conducted the syphilis study held the preconceived notion that African Americans were predisposed to sexually transmitted diseases, without any medical evidence to prove this belief. But as Fullilove (1998) argues, we need only to consider the observation that sexually transmitted diseases, which are overrepresented among African Americans and Hispanics, can be best understood by examining the geographically organized social networks within which they are spread. Rather than a narrow focus on a nonmedical myth based on racism, researchers in the Tuskegee Syphilis Study should have been focused on the social and medical factors that may have caused an increased number of syphilis cases among African Americans in Tuskegee, Alabama, and its surrounding areas. Could it have been the lack of access to medical care or even a lack of exposure to health care education resources that caused increased risks of sexually transmitted diseases among these susceptible populations? This question and other such pertinent questions should have been explored by public health officials who conducted the Tuskegee Syphilis Study.

Race Classification:
Colorism within the Tuskegee Community

Colorism, the preference or prejudice shown to people of color depending on the lightness or darkness of their skin, has been historically and systematically ingrained within the black community since the early days of North American slavery (Hare 2010). In a 2005 working paper by the National Bureau of Economic Research, the history of colorism was examined. The study found that colorism affected the household wealth of fifteen thousand households during the 1860 census. For example, in the urban South, researchers found major differences in wealth among white, biracial, and black households (Hare 2010). The paper notes: "For many of African descent, black is not black—both in terms of how they view themselves and how others view them. There are meaningful subtleties of shade, differences that have socially and politically meaningful distinctions today—just as they have in the past" (National Bureau of Economic Research 2005, 1).

The paper entitled, "Colorism and Wealth," confirmed that while individuals tend to define people as either black or white, such a dichotomy ignores the children produced from white slave owners and their black slaves (National Bureau of Economic Research 2005). The paper also reported that people of color with lighter skin received preferential treatment during slavery from slave owners, such as working in the home instead of the fields (National Bureau of Economic Research 2005). As Radhika Parameswaran, an associate professor of journalism at Indiana University, explained: "Colorism within the black community became about class or associations with whiteness" (Hare 2010, 1). The paper cites that slave owners often included their light-skinned house slaves in their wills. Mulattoes also had greater access to education and better jobs, especially in the antebellum South (National Bureau of Economic Research 2005).

ISSUES OF RACE, JUSTICE, GLOBAL COLORISM, SKIN BLEACHING, AND INTERNALIZED PERCEPTIONS OF SKIN COLOR

The preferences shown toward the mulatto population of blacks continued even after the American Civil War. Research studies on color classification were conducted during the 1930s and 1940s, which found that "light skin tones and perceptible traces of non-African heritage were associated with material advantages for African Americans" (National Bureau of Economic Research 2005, 1). The paper reported that a 2000 analysis indicated that light-skinned men were twice as likely to have received high prestige employment opportunities than their dark-skinned counterparts (2005). This factual data further emphasizes how ingrained cultural norms of race affected the selection of the educated and upwardly mobile mulatto elite class and the formative increase of the rural darker-skinned poor, creating a distant divide between the haves and have-nots according to skin color. As previously mentioned, it is no wonder an experiment such as the Tuskegee Syphilis Study was allowed to be sustained through medical deception promoted by the US Public Service and the mulatto elite of Tuskegee Institute.

To be true to one's calling, it is a requirement to expose, rather than hide, the hard moral and ethical truths of racism in the sciences, even if this truth offends one's own race. With that said, internal aspects of race classifications have stigmatized and marginalized African Americans of the brown and darker hue. In many instances, opportunities for economic, social, and professional advancement within the African American community have not been based on merit and talent but rather historically based on the tonality of one's skin. No other example of historic class, economic, educational, and skin color discrimination could be more evident than the city in which the Tuskegee Syphilis Study took place—Tuskegee Institute, Alabama. Housed in a small, rural community, Tuskegee, Alabama, was once a town literally split in two—Tuskegee Institute (mostly where black staff, faculty, and administrators of the college lived) and Tuskegee (strictly segregated to include the white citizens of the community). Tuskegee Institute was first led by a mulatto principal named Booker T. Washington and established in 1881 through the political ingenuity of another mulatto, Lewis Adams, a distinguished community leader and businessman.

With lighter skin came respect and approval within the segregated white and black communities of Tuskegee, Alabama. As evidenced by Principal Washington in his autobiography, *Up from Slavery*, when Tuskegee elite community leaders wrote Hampton University's principal, General Samuel C. Armstrong, they requested a white male principal, as was the norm in the 1800s. But when General Armstrong finally responded to their request, he stated that he only had a mulatto young male graduate of Hampton University who was highly qualified. Once the community leaders from Tuskegee received General Armstrong's response, they asked for Booker T. Washington to come at once (Washington 1901). And the rest, essentially, became history.

As evident in appendix A, the educational setting of Tuskegee Institute consisted of faculty members who were mostly of lighter complexion because many were educated at northern white colleges where they "passed for white" or were direct descendants of their white slave masters. Access to higher education for many light-skinned African American former slaves was more plentiful because of their direct access to their white counterparts' educational resources, such as books. In most respects, lighter-skinned slaves were viewed as less threatening and visibly more similar in skin color to their white counterparts than darker-skinned slaves. As such, these lighter-skinned slaves were known as *mulattoes* due to their light skin color and the likelihood that they were born as a result of interracial sexual intercourse with or rape by the white slave masters or mistresses. These slaves were known as *house Negros*, who worked in their slave master's home and often played alongside the slave master's children. Oftentimes, the house Negro mulatto children were half brothers or sisters to their white counterparts due to miscegenation.

The slaves who worked the planation fields were known as *field Negros*. They tended to be of a darker skin complexion and were not provided many of the privileges house slaves were given. They worked the plantation land for such products as cotton. As Malcolm X described, "The house Negro usually lived close to his master. He dressed like his master. He wore his master's secondhand clothes. He ate food that his master left on the table. And he lived in his master's house—probably in the basement or the attic—but he still lived in the master's house" (1963, 1). Once slavery was formally abolished through the Thirteenth Amendment to the United States Constitution in 1865, many of the field slaves continued to work the land as farmers in order to make a living for their families. This scenario was most evident in Tuskegee, Alabama, where many field Negroes essentially became the land farmers who were direct descendants of an enslaved people.

As educational opportunities for African Americans expanded in Tuskegee, Alabama, due to the establishment of Tuskegee University, those who were the most notable educators and administrators were also the lightest in skin complexion, as evident in faculty and staff photographs with Principal Booker T. Washington and subsequent early presidents of the institution. In effect, lighter-skinned African Americans historically worked as prominent staff, faculty, and administration officers in the ivory tower system of Tuskegee Institute, while many brown and darker-skinned African Americans served as land farmers in Tuskegee, Alabama, and surrounding areas (Carmichael 1966; Frazier 1957; Russell, Wilson, and Hall 1992). Thus, the systematic foundation of slavery's house and field Negro approach to class and race distinctions has persisted and has been perpetuated through internalized racism within local communities, such as Tuskegee Institute, Alabama (X 1963).

Nonetheless, most African American doctors on staff at the John A. Andrew Memorial Hospital (the primary Tuskegee University campus facility where the syphilis study was conducted) were among the middle class of Tuskegee (Jones 1993). The African American farmers involved in the syphilis study were mostly from the rural community in and around Tuskegee Institute, Alabama (1993). They were considered by most doctors involved in the syphilis study as simply research subjects, not human beings who deserved the right to their own self-determination or autonomy to make medical decisions (1993). Thus, it is no wonder that as a consequence, internal race delineations within the early African American community of Tuskegee may have led to the unconscionable—the institutional and administrative approval of a study funded by the US Public Health Service based on racial myths rather than scientific facts.

More concrete public health questions that researchers in the syphilis study should have been concerned with include the following: How do factors, such as the lack of access to health education and health care lead to disproportionate cases of syphilis among African Americans within a Black Belt community? Rather than a medical focus on the cause of syphilis within the African American community in the Tuskegee region, medical researchers were preoccupied by racist and scientifically unfounded misconceptions about the black male body, which falsely theorized that African Americans by nature were more predisposed to contact venereal diseases than any other human race. A more appropriate study would have focused on the effects of racism on health disparities rather than a use of race as a qualifier to systematically withhold treatment for syphilis when penicillin became widely available during the height of the study. As Krieger (2003) points out:

> Skin color, in turn, would become a racialized expression of biology if, absent any evidence, it were treated as a valid

marker for other unspecified genetic traits, reflecting a presumption that the biology of "race" equals the biology of gene frequencies, This was the logic of the flawed research agenda egregiously exemplified by the Tuskegee syphilis study, unnaturally intended to determine whether the "natural history" of untreated syphilis in Blacks was the same as that previously observed in Whites, in light of hypothesized differences in their nervous systems. (195)

During the Tuskegee Institute's most fertile days, the categorization of skin color was a significant factor in a group or person's social and even economic standing in American society. Social and political powers were thus defined by race and one's association or disassociation with the black race. Dr. Eugene H. Dibble, Jr., medical director of the John A. Andrew Memorial Hospital, along with the principal of Tuskegee Institute, Robert R. Moton, clearly understood the nature of the experiment and gave their support to it (Brandt 1978). It is worth noting that Dr. Dibble, an American mulatto, and his family were members of the upper middle class black elite in America (Graham 1999) (see appendix B).

Therein lies a debate of class difference in how Dr. Dibble valued and viewed other black men of the darker race who were going to be subjected to nontreatment within the experiment. Most doctors involved in the study, including Dr. Dibble, simply viewed the black men experimented upon as only research subjects, a means to an end, not autonomous human beings worthy of consent for their research participation (Thomas and Quinn 2000). The wealth of mulatto black professionals was highly maintained and further increased because health officials, such as Dr. Dibble, were paid an additional salary by the US Public Health Service (USPHS) in order to convince the black male participants in the experiment that these black officials had capital and validity in the eyes of the federal government. In turn, the USPHS gave Dr. Dibble, the medical director of the hospital at Tuskegee Institute, a paid interim appointment to the US Public Health Service (Brandt 1978). As Dr. O. C. Wenger, chief of the federally operated venereal disease clinic at Hot Springs, Arkansas, noted:

> One thing is certain. The only way we are going to get postmortems is to have the demise take place in Dibble's hospital and when these colored folks are told that Doctor Dibble is now a Government doctor too they will have more confidence (NA-WNRC 1933, August 5, 1).

As awareness of the destructive stages of syphilis increased in medical practice, many black males in the study remained untreated for syphilis even until death. The agreement among university officials and the US Public Health Service was that syphilis would not be treated among the population of black men until they reached autopsy in order to test the scientific hypothesis that the black male body physically reacted to syphilis differently than whites. As Surgeon General H. S. Cummings explained in a letter to Principal Robert R. Moton of Tuskegee Institute:

> This study which was predominantly clinical in character points to the frequent occurrence of severe complications involving the various vital organs of the body and indicates that syphilis as a disease does a great deal of damage. Since clinical observations are not considered final in the medical world, it is our desire to continue observation on the cases selected for the recent study and if possible to bring a percentage of these cases to autopsy so that pathological confirmation may be made of the disease processes (NA-WNRC 1933, July 27, 1).

ISSUES OF RACE, JUSTICE, GLOBAL COLORISM, SKIN BLEACHING, AND INTERNALIZED PERCEPTIONS OF SKIN COLOR

An appropriate metaphorical synopsis of the relationship among black officials at Tuskegee Institute, the research study subjects, and the US Public Health Service can be summarized in the prophetic words of race thinker, Malcolm X (1963):

> If the master's house caught on fire, the house Negro would fight harder to put the blaze out than the master would. If the master got sick, the house Negro would say, "What's the matter, boss, we sick?" We sick! He identified himself with his master, more than his master identified with himself. And if you came to the house Negro and said, "Let's run away, let's escape, let's separate," the house Negro would look at you and say, "Man, you crazy. What you mean, separate? Where is there a better house than this? Where can I wear better clothes than this? Where can I eat better food than this?" That was that house Negro. In those days he was called a "house nigger." And that's what we call them today, because we've still got some house niggers running around here.
>
> This modern house Negro loves his master. He wants to live near him. He'll pay three times as much as the house is worth just to live near his master, and then brag about "I'm the only Negro out here." "I'm the only one on my job." "I'm the only one in this school." You're nothing but a house Negro. And if someone comes to you right now and says, "Let's separate," you say the same thing that the house Negro said on the plantation. "What you mean, separate? From America, this good white man? Where you going to get a better job than you get here?" I mean, this is what you say. "I ain't left nothing in Africa," that's what you say. Why, you left your mind in Africa. (1)

In summary, the relationship among the upper-middle-class mulatto medical staff of John A. Andrew Memorial Hospital at Tuskegee Institute, the darker black male farmers who served as research subjects in the study, and the US Public Health Service was metaphorically comparable to the house and field Negros' relationship with their slave master.

The US Bureau of Census and Race Classification Structures

The US Bureau of Census played a pivotal role in defining race and in how African Americans viewed themselves as being mulatto (blacks "mixed" in racial terms with white origins) or black (of pure African ancestry and of the "darker race"). This racial order or classification system helped define the haves and the have nots, the educated and the uneducated, and the social elite and lower classes within the black community. As Hochschild and Powell eloquently argue, census classification was "primarily a mechanism for racial control and exclusion—an instance of the broader claim that classification is part of the construction and maintenance of racial hierarchy and the distinction between insiders and outsiders" (2008, 9).

As a result of these racial order classifications within the black community, many black Americans who were not racially mixed with white blood were ostracized, isolated, and labeled as "unfit" to obtain the American dream, while those of mixed or mulatto heritage were given greater economic, educational, political, and social access. In turn, much of what was labeled as *scientific inquiry* through experiments, such as the Tuskegee Syphilis Study, were simply masked forms of negative eugenics practiced on the darker race of black men in order to scientifically prove that the darker race of black Americans were biologically inferior.

ISSUES OF RACE, JUSTICE, GLOBAL COLORISM, SKIN BLEACHING, AND INTERNALIZED PERCEPTIONS OF SKIN COLOR

Before the Tuskegee Syphilis Study (1932–1972), the US Bureau of Census contributed to race classifications within the black community between 1850 and 1930. The study was conducted in the South during America's most racially charged era of social, educational, political, and economical segregation. Historically speaking, the US Bureau of Census essentially conspired with southern politicians to produce a census policy based on race. Nobles (2000) notes: "The former sought to generate data supporting their theories of white biological superiority, while the latter sought evidence to justify and legitimate policies of non-white exclusion and white supremacy" (9).

As previously noted, the Tuskegee Syphilis Study was without question a race-based study. Because dark-skinned black men were viewed as genetically inferior and lighter-skinned blacks were considered more culturally assimilative and closer to white racial purity, many black men who were research subjects in the syphilis study were field farmers of the darker male race. This race classification structure was historically rooted in the systematic practice of allowing light-skinned Negro slaves to work in the slave master's home, while forcing dark-skinned slaves to work in the fields. As a result, the dark skin color of the research subjects in the syphilis study was used as a means to justify race-based research practices, which emphasized that darker-skinned blacks were genetically inferior to whites, and those associated with whiteness, such as mulattoes, were genetically superior (see appendix C). As one television newsmagazine confirmed and reported, the research subjects recruited for the Tuskegee Syphilis Study were direct descendants of slaves from the Deep South of Alabama (*Prime Time Live*, 1992). But more curiously, these farmers were visually of the darker race of blacks, which is consistent with the imagery of black field Negro slaves who worked the cotton fields before emancipation.

As a result, the selection and recruitment of dark-skinned black men in the Tuskegee Syphilis Study was no accident. These men were selected based on the social, scientific, economic, political, and educational structures rooted in segregation practices among black and white race classification systems within American society. It is no wonder that rigid system nuances in racial classifications naturally transferred soon after slavery into health and science practices in medicine through eugenically and racially biased studies such as the Tuskegee Syphilis Study. Thus, "politics, science, and ideology were inextricably mixed" (Hochschild and Powell 2008, 15).

From the syphilis study's very beginning, scientific exploration of racial mixtures within the black community was greatly studied and explored. E. Franklin Frazier (1933) and other research scholars previously pointed out, "scholarly attention to racial mixture continued into the 1930s, and mixture never stopped being socially and economically salient in black communities" (19). Much of the racially charged data gathered about mulattoes and darker-skinned black Americans from the US Bureau of Census solidified existent myths about darker-skinned black Americans and reemphasized internalized cultural prejudices within the black community that mulattoes were inherently "better" due to their racial proximity to whiteness. Hochschild and Powell (2008) signal that US Census authorities "presumably wanted to know how much the racial hierarchy needed to be further reinforced, which depended in part on whether the number of people blurring the line between the subordinated and the dominant races was increasing or decreasing, thriving or struggling, moving toward or away from whiteness" (16).

Sociologist W. E. B. DuBois, himself a product of the mulatto class, recognized the detrimental effects of race classifications within the black community. DuBois, along with a number of light-skinned black leaders, including Booker T. Washington, called for an end to rigid distinctions of color within the black community. These leaders claimed that distinguishing darker-skinned blacks from mulattoes was conceptually misguided and destructive to political, educational, and economical solidarity (Hochschild and Powell 2008). In fact, DuBois (1900) urged US Census officials to "class those of African descent together" when surveying mulattoes and darker-skinned black Americans (307). Although scientific inquiry into racial mixture has shifted, the codification of class classifications has persisted within the African American community. It continues to permeate the educational, medical, social, political, and economic tapestry of the black community.

Justice
Justice as Defined Historically through the Belmont Report

In 1979, the first version of the *Belmont Report* was published seven years after unethical medical actions conducted in the Tuskegee Syphilis Study were revealed to the general public.

The report offered ethical principles and guidelines for the protection of human subjects in research. These principles included respect for persons, beneficence, and justice. For the framing of my paper, I plan to expound on the principle of justice as it relates to the Tuskegee Syphilis Study. In the context of ethics, justice may question who ought to receive the benefits of research and eventually bear its burdens. According to the *Belmont Report* (1979), justice comments on the fairness of distribution or what is deserved. In turn, the principle of justice connotes that equals ought to be treated equally (*Belmont Report*, 1979). But for our discussion here, how do we measure equality? What is equality based upon? As the *Belmont Report* states: "An injustice occurs when some benefit to which a person is entitled is denied without good reason or when some burden is imposed unduly" (1979, 4).

As the *Belmont Report* (1979) notes, historically, experimental research among research subjects largely was conducted on poor ward patients during the nineteenth and twentieth centuries. The research results from these studies usually benefitted privileged patients. This brings to bear the grave medical research disparity among the poor and disenfranchised. The Tuskegee Syphilis Study is a stark example of how research experimentation lacked equal distribution of patient participation among class stratospheres. Thus, the experiments at Tuskegee reflect the lack of distributive justice in clinical research. For example, the *Belmont Report* (1979) reflects that "the Tuskegee Syphilis Study used disadvantaged, rural black men to study the untreated course of a disease that is by no means confined to that population. These subjects were deprived of demonstrably effective treatment in order not to interrupt the project, long after such treatment became generally available" (5). As an example of the lack of justice, the selection of research subjects in the syphilis study was not critically scrutinized. Therefore, some classes of individuals were systematically selected because of their availability and were ethically compromised, medically deceived, and manipulated in order to participate in the study.

The concept of burdens and benefits also plays a pivotal role in the analysis of medical justice. The burden of participating in the syphilis study for the men involved far outweighed the benefits of research participation because the research subjects did not receive treatment for syphilis even when penicillin was available to treat the disease. As a consequence, many men died as a result of nontreatment of syphilis, which constitutes a form of ethical injustice in medical treatment and health-care research.

ISSUES OF RACE, JUSTICE, GLOBAL COLORISM, SKIN BLEACHING, AND INTERNALIZED PERCEPTIONS OF SKIN COLOR

Two Philosophical Concepts of Justice according to John Rawls and John Locke

Philosophical thinkers, John Locke and John Rawls, have hinted at solutions to health care and research inequalities in writings centered on justice, even to the extent of reparations for victims. This brings to mind two fundamental questions: Do such populations deserve reparations for crimes committed against them by public health care physicians? Should these same officials be criminally and ethically accountable for their unjust and unethical medical actions against vulnerable populations? Modern philosopher John Rawls has helped me explore these questions through his theory of justice as fairness that can also be applied to issues of inequality and inequity in health care research.

By most accounts, John Rawls was considered one of the most well-known theorists on the philosophical concept of justice. He believed that a just and equitable health care system preserves the right and ability of people to participate in the social, political, and economic life of society (Daniels 2012). Rawls's theory of justice as fairness assumes that we have a completely healthy society. He argued that "a just society must assure people equal basic liberties, guarantee that the right of political participation has roughly equal value for all, provide a robust form of equal opportunity, and limit inequalities to those that benefit the least advantaged" (2012, 11). Rawls concludes that once these requirements of justice are met, we, as a society, can be assured that respect for self-worth is visible and evident. The fair terms of cooperation identified through these principles eventually promote our political and social well-being (2012). If the least among us is provided with equal social necessities such as fairness and justice in health care, our collective population benefits politically and socially because society is better off with a healthier people.

The basic tenants of Rawls's theory of justice as fairness (equal liberties, equal opportunity, a fair distribution of resources, and acknowledgement of our self-respect) all contribute to the elimination of injustice in health care research practices (Daniels 2012). Therefore, social justice is inextricably connected to justice as a means to fairness in medicine, health care, and research participation. For instance, the equal and fair distribution of health-care resources sustains all people as full and engaged citizens. Equitable health care also maintains the normal functioning of individuals, protecting a person's fair share of opportunities to succeed (2012). In terms of recruiting human subjects, researchers should seek to protect all participates in human experimentation, regardless of race, ethnicity, gender, sexual orientation, religion, or any other discriminatory factor. Researchers must also attempt to represent a cross section of the human race in their research sample when appropriate. Such research quality actions help assure vulnerable populations that they are valued equally in research while fairly being recruited and selected as research participants.

Rawls's theory "assures people of equal basic liberties, including equal access to political participation; guarantees a robust form of equal opportunity; and imposes significant constraints on inequalities" (Daniels, Kennedy, and Kawachi 1999, 217). Democratic equality is the central outcome. It ensures that all citizens have the social bases of self-respect with the conviction that prospects in life are fair (1999). The concept of justice as fairness is unique to the discussion at hand because it explicitly supports the notion that equal opportunity can also be extended to medical research studies.

In many ways, our health is affected by our social position or status and the underlying inequalities in our society (Daniels 2012). Who we are in relation to our socioeconomic background, race, ethnicity, regional location, and gender are just a few factors that affect fairness and justice in research. In order to formulate a fairer and more just health-care research system, we must also consider social justice measures that will prevent unethical research studies such as the experiments at Tuskegee.

ISSUES OF RACE, JUSTICE, GLOBAL COLORISM, SKIN BLEACHING, AND INTERNALIZED PERCEPTIONS OF SKIN COLOR

The philosophical theory of justice as fairness serves as an effective model that can help reduce unjust clinical research practices on vulnerable populations. One explainable reason why human subjects involved in the experiments at Tuskegee were recruited and medically exploited was because they lacked access to equalized health care resources in their rural communities. These participants were primarily interested in participating in the clinical trial because public health doctors assured them that they would be treated free of charge for their *bad blood*, a term commonly used in their local community to describe people infected with syphilis.

In turn, public health officials distributed flyers throughout the local community and recruited men from the black church in order to participate in the study. These men were selected and recruited to participate in the study because of their race rather than a quest for true scientific discovery. Because of the public segregation of hospital facilities in the South during the 1930s, the closest and most accessible hospital for nonmilitary black males was the John A. Andrew Memorial Hospital. This hospital was located on the campus of Tuskegee Institute and served as the main experimental site for the Tuskegee Syphilis Study. If limited health-care education resources and the segregation of hospital services were not prevalent factors during the occurrence of the syphilis study, many black men would not have participated in this clinical trial. Instead, they would have sought out or been provided equalized health education and medical treatment for syphilis.

In order to move forward in our discussion of John Locke's philosophical emphasis on reparations and retributions, it would be practical to explain a basic attribute to his theory: retribution. Retribution is the balancing of a wrong through punishment (Pollock 2005). In a social contract, the state or governing institution has the right to inflict retribution or punishment upon guilty parties or criminals who transgress against the individual rights of others. This concept is as ancient as some forms of philosophy, which is also evident in the writings of Thomas Hobbs (*Leviathan*, 1651), Jean-Jacques Rousseau (*Du contrat social*, 1762), and the philosopher of focus in this essay, John Locke (*Two Treatises on Government*, 1690).

In Locke's *Two Treatises on Government* (1690), he acknowledges that all men are created equal in a state of nature by God. Thomas Jefferson, an American founding father, adopted this belief and infused it within the US Declaration of Independence. Such a belief holds that by nature, each person possesses the right to life, liberty, and the pursuit of happiness (or more directly, property). Those who hinder these fundamental rights should be held accountable to society.

As we observe the blatant disregard for human rights as demonstrated in the experiments at Tuskegee, we see how individual rights to life and liberty were violated and disregarded. This purposeful lack of informing human subjects about their medical status and the full extent of invasive medical procedures goes against the medical ethic to *do no harm*. It is the duty of doctors to work in the best interests of patients rather than to work against the formal or informal ethical standards that exist in medical practice. Locke would clearly argue that the experiments at Tuskegee violated the natural rights of research participants. He would also argue for reparations to citizens based upon the human suffering of their individual participation.

In chapter 2 (sections 8 and 11) of *Two Treatises on Government*, Locke (1690) advocates for the punishment of crimes committed against society that violate the rights of others. He also further indicates that reparations are viable options that may help restore violated liberties back to victims, restrain others from committing unlawful acts, and deter against criminal acts in society. One could justifiably extend Locke's concept of reparations to include unethical and unjust medical acts against human subjects who have been stripped of their ethical and human rights.

Therefore, reparations for victims or family members affected by the Tuskegee Syphilis Study are warranted. In this case, reparations would help repair past medical wrongdoings, deter others from conducting such unethical acts, and help heal the physical, mental, and spiritual scars of injustice. Therefore, reparations can occur in various forms from monetary to educational resources that have, otherwise, been limited to vulnerable populations.

The Nature and Purpose of the Institutional Review Board (IRB) as a Means to Enhance Distributive Justice in Human Experimentation

Ethical and justice-based regulations governing the protection of human subjects became effective in 1974 by the establishment of institutional review boards (IRBs). IRBs were developed as a response to unethical experimentations, such as the Tuskegee Syphilis Study. After the study ended in 1972, it was determined that the human subjects involved were not properly informed of the true nature and extent of the clinical trial. Today, the role of IRBs is to serve as the primary mechanism by which research participants are protected (Greene and Geiger 2006). According to Klitzman (2013), IRBs also review research proposals (approve, suggest specific revisions, or disapprove these proposals); provide continuing review and approval of research projects (at least annually); conduct in-service training about human participant protection in research; respond to consumer inquires and concerns about their participation in research; and administer technical assistance to researchers. Most significantly, the main function of IRBs is to protect human subjects in research based on ethical rules, regulations, and policies.

The actual review processes and philosophies vary from institution to institution based on local interpretation and application of federal regulations. Such flexibility is appropriate because of the unique considerations in research environments locally (Greene and Geiger 2006). Therefore, IRB approval is locally based because of the local committee's specific and unique understanding of the ethical challenges and issues related to the clinical trial being reviewed. In addition, IRBs are institutionally based because they understand the local context, population, and issues involved in the proposed research. These safeguards allow IRBs to function locally in order to protect its citizens. Some recommendations to improve IRBs and increase distributive justice in human subjects research include sponsoring professional development and educational opportunities for committee members in order to meet the growing religious, philosophical, and ethnic diversity of some local communities.

Informed Consent as a Tool to Enhance Distributive Justice in Human Experimentation

Several strategies can help enhance distributive justice in human subjects research by effectively implementing informed consent. For example, as indicated in display 1, distributive justice can be enhanced by researchers developing effective informed consent documents and good decision making practiced by study candidates.

Display 1

Distributive Justice Enhancement Model

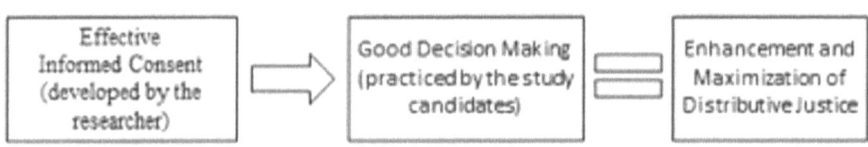

ISSUES OF RACE, JUSTICE, GLOBAL COLORISM, SKIN BLEACHING, AND INTERNALIZED PERCEPTIONS OF SKIN COLOR

The American Medical Association (2009) defines informed consent as the "process of communication between a patient and physician that results in the patient's authorization or agreement to undergo a specific medical intervention." As Mackintosh and Molloy (2003) point out, improving the informed consent process and consent forms help subjects better understand clinical research and reduce the number of invalid consents obtained. As a policy, institutional review boards (IRBs) usually require that consent forms be written at a sixth- or eighth-grade level (Hochhauser 2004). But according to Hochhauser, more research on the consent process and a further development of strategies to improve subject understanding is needed to improve consent documents (2004). Levine (1984) recommends that procedures be introduced to enhance the process of informed consent by making complex information more comprehensible to the prospective research participant.

Another vital tool to increase distributive justice is through effective informed consent practices, such as eliminating confusing and deceptive medical metaphors and/or any language that confuses, omits, or distorts vital information. Informed consent forms should be more specific and less complex in use of language. Researchers should also be aware of using metaphors that may deceive potential research participants. For instance, when public health doctors knew that the term *bad blood* was not an accurate description of syphilis, they medically deceived the study participants to believe that they would be cured of their *bad blood* by participating in the experiments at Tuskegee. Research investigators should avoid any usage of language that confuses or possibly deceives research candidates. The simple devotion of more time for explanations, use of local language, or obtaining consent in home environments have also improved the informed consent process (Sarkar et al. 2010) while reducing the chances of inequality and injustice.

In rural areas like Tuskegee where health care resources may be limited, educational tools can be designed to improve understanding and increase the likelihood of receiving genuine informed consent from research participants (Gazzinelli et al. 2010). In many cases, special educational videos can be produced to explain the content and procedure of clinical trials in order to assist in the informed consent process (2010). Visual aids, such as pictures, can also be included in informed consent documents to aid comprehension and limit instances of possible unjust medical treatment. Sarkar et al. (2010) has also noted that "using a simplified version of the written consent document with pictorial representation and the use of consent advocators or professional nurses with prior research experience have also been shown to improve the participants' comprehension" (4).

Other authors have commented on how to improve patient comprehension in informed consent. Valid informed consent requires that patients understand the proposed intervention, including potential risks, benefits, and medical alternatives. The inability to obtain informed consent can compromise patient autonomy, place the patient's safety at risk, and legally may constitute negligence or battery (Schenker et al. 2011). The researchers found that initiatives such as additional written information, audiovisual/multimedia programs, extended discussions, and test/feedback techniques improve patient comprehension in informed consent, which eventually promotes a fairer and more equitable mechanism for participation in medical research (2011).

Race-Based Justice
Conceptual Definition and Discussion
Does Race-Based Justice Exist? And Should Justice Be Race-Based?

Justice based on race for Tuskegee human subjects and their families cannot be fully accomplished without certain elements that are pertinent to the African Americans who were affected by these clinical experiments at Tuskegee. There are two central questions we must discuss in order to further explain race-based justice: 1) Does race-based justice exist? 2) Should justice be race-based? Race-based justice does exist, but it must be inclusive of spiritualism centered within the black church because many human subjects were recruited from this setting. But, on the other hand, justice, in and of itself, should not be racially based. For the discussions ahead, let us first offer a definition of race-based justice. Since the concept of race-based justice is an evolving ideal, I offer a working definition based upon the justice-based teachings of theorist John Rawls. Therefore, race-based justice may be viewed as the principled allocation of resources among a racial population as a result of previous unethical and immoral wrongdoings based on the race of a particular group of people.

Although race, in and of itself, is a sociological and sometimes generic abstract categorization, it impacts the social, economic, educational, political, medical, and religious facets of populations who are mistreated based on race. As previously noted, because many men who participated in the Tuskegee Syphilis Study were recruited from the black church, it is crucial that the spiritual component of African American religious culture be integrated within a race-based justice theory. Within the essence of the black church, it is paramount that issues of justice and race are interconnected to spirituality. In other words, because of the nature of the black church and its founding, it is very difficult to separate the nature of the spiritual being from one's race and issues surrounding justice.

Asian American racial justice theorist Harry Chang viewed race from a multilayered perspective. Influenced by the writings of Karl Marx and other philosophers, Chang's work laid the foundation of two pivotal theories of racism: the theory of racial formation and critical race theory. During the 1970s, his ideas were highly controversial because they sparked uncomfortable conversations about race (Wing 2007). Chang highlighted the invalidity of the "one-drop" rule that determines race in the United States. By shedding light on this rule, he showed that racial categories are sociohistorical categories, not genetically or genealogically based (Wing 2007). Moreover, according to Chang, racial categories are qualitatively distinct from class, ethnicity, or nationality categories. Moreover, Chang coined the term *racial formation* in order "to underscore the necessity of analyzing racism as a historical process that encompasses the origins of racism, how and why it has changed over time, and the process of eliminating it in a given historical context" (Wing 2007, 1). In addition, Chang argued for the centrality of law to racial formation and the inseparability of race and class. His clarification of the distinctiveness of racism laid the foundation for analyzing the intersection between race and nationality (2007).

Chang believed that the works of Karl Marx, especially *Capital*, could be applied to understanding racism in North America. As Chang (1973) so eloquently observed in "Racism and Racial Categories":

> More than half of *Capital* is devoted to the critique of bourgeois political economic categories [commodity, value, money, capital, etc.]. It is a key insight of historical materialism that historical development is reflected in the logical development [the development of concepts and categories] and, as Engels put it, the latter, as a result, represent the former "in complete maturity and classical form." Hence the critique of capitalist categories plays a crucial role in the analysis of the capitalist mode of production. This is also the method that must be applied to the race question. A Marxist analysis of racism must begin with a critique of the racial categories (black and white) themselves, and from there proceed to an examination of the

socio-historical situation that endowed these forms of thought with deadly social validity. (1)

The theoretical insights of Chang's racial justice theory concluded that racism itself should only be the subject, not the object of study. According to Chang, the object of study should be racial categories and the social practice that produced them (Wing 2007). Therefore, he sought to overcome the subjectivity in race theorizing, especially in relation to the transposition of the race question into an internal colonial, national, and/or ethnic question (Wing 2007). Chang noted: "The dialectic of the categories of Black and White is relative as opposed to absolute" (2007). This means that, like class categories (workers versus capitalists, landlords versus peasants, etc.), categories of black and white are determined by and only in relation to each other. Therefore, the categories of "white" and "black" are not eternal, independent, or neutral categories (Chang 2007). They are the necessary ideological representations of the mutually dependent and exclusive poles created by the social relationship of racism. For example, US slavery was the only modern slave society at the time that solely depended on slave reproduction rather than the African slave trade for most of its labor force (2007).

As Marx had concluded that the formation of money was linked to commodity production, Chang discovered that racial formation in the United States was linked to the historical conditions of the development of capitalism (2007). His study of the historical origins of racism in the colonial United States led to the discovery of the relationship between the development of racial categories and the development of the slavery-based plantation system (Wing 2007). He noted that "racism was not a mere 'add on' to US capitalism, but a central condition without which US capitalist development would have been qualitatively different. In other words, racism could powerfully shape capitalism, not just vice versa. Racism did not just add additional profits to capital, it actually determined the very shape and course of capitalist accumulation, social formation, and political structure in the United States" (2007, 1).

The work of Chang highlighted that the "planter" or "slaveholder" was not a single economic role but in fact involved in a variety of economic relationships, such as landowner (or tenant), slave owner, slave user (or renter), slave buyer/seller, and banker (or debtor) (Wing 2007). Most important, as Chang (1974) revealed, "The distinction between slave-owning and slave-using corresponds to a case of capitalistic dialectic of distribution and production" (1). The historical circumstance of understanding the origin of slavery in North America leads one to conclude that slave labor originated as a substitute for wage labor in a system of production demanding wage labor but short of such labor. "Thus, although slaves in the US were not wage-earners, they were the labor counterpart of capital with this proviso: the planter expended wage in his capacity of agricultural capitalist and intercepted wage in his capacity of capitalist slave owner" (1974, 1). Chang's philosophical concepts of race helped shape the evolution of race justice theory, which emphasized principled and equal justice based on historical understandings of racial injustice perpetuated by a majority population. Therefore, in order to fairly provide justice to an oppressed people, issues of racial discrimination and subjugation must be reconciled.

Black Soul Repair (BSR): A Holistic Model for Restorative Justice

In order to provide a tangible framework by which the theoretical concept of race-based justice can be translated, I developed the Black Soul Repair (BSR) Holistic Model for Restorative Justice (see display 2). Moreover, the model serves as a practical means by which race and justice can be implemented. It is based on the notion of restorative justice, the belief that essentially means to restore or bring back together to the whole of personhood through just, fair, and equitable means (Prison Fellowship International Centre for Justice and Reconciliation 2013). My model's circular form is symbolic of the belief that the three fundamental elements of healing are inclusive, interconnected, and essential to holistic healing. These three ethical tenants of healing consist of *spiritual, physical,* and *mental elements*.

Display 2:

Black Soul Repair (BSR)
A Holistic Model for Restorative Justice: *Healing* the *Whole* of a *Community* in *Pain*

Spiritual healing is inclusive of the strategic involvement of African and ancestral spiritualism and the black church. The black church, especially, previously played a significant role in recruiting many clinical trial subjects, including the recruits in experiments at Tuskegee (Jones 1993). Therefore, it is my belief that they must also play a prominent role in restoring truth and reconciliation among vulnerable black populations that need to be involved in current clinical trials of great importance.

Physical healing consists of preventive medical measures, such as early detection and the increased involvement in clinical trial research that affects the black community. If patients are aware of the risks and benefits involved in nontreatment and early detection, then they are more likely to seek available means of treatment and care. In effect, patients would also be more willing to participate in clinical trials that are of benefit to themselves and their community if they understand the ramifications of human subjects research participation.

Mental healing is cognizant of the act involved in changing communal perspectives on issues, such as internalized colorism, melanin, clinical depression, and medical mistrust in order to restore and renew self-confidence and self-efficacy. Although the Black Soul Repair Holistic Model for Restorative Justice focuses on healing strategies for families affected by the Tuskegee Syphilis Study, this model is also applicable to other victims and their families who have been affected by unethical clinical studies held domestically and internationally.

Conclusions
Ethical Recommendations for Solutions to Race Challenges in Biomedical Research

Racial consciousness has been ingrained within American culture for decades. Since the founding of this nation, one racial population has dominated the power structure while other vulnerable groups have been underserved, especially in health care and education. These subversive acts of inequality have led to racial disparities economically, politically, socially, educationally, and medically. But how can we, as a nation, repair, resolve, and heal from the health disparities that damage all people? What can we do as public health practitioners to help alleviate racial injustice in health care, which has resulted in many minority populations lacking basic health care? In the closing section of my thesis, I plan to provide underutilized solutions to race challenges in health care in light of how the experiments at Tuskegee can serve as a case study for ethical improvement and sustainability. The areas of my focus include the establishment of a US Bioethics Commission on Truth and Reconciliation, the Concept of Ubuntu, an apology for racially based injustices in health care research, clinical empathy, and education. At this moment in our discussion, I now focus on an establishment of a US Bioethics Commission on Truth and Reconciliation as a viable mechanism to help alleviate injustice and racially based biases in health care.

US Bioethics Commission on Truth and Reconciliation

Reconciliation is an unexplored area in bioethics that can help resolve racial issues in medicine and health care. Within my discussion of race and justice challenges in clinical research, reconciliation and healing play integral roles in bringing forth equality for all. Truth-telling and an inclusion of stories by victims and perpetrators of injustice should be considered as vital weapons to extinguish injustice in medical research similar to the ethical tenants of the historic Truth and Reconciliation Commission of South Africa. The Truth and Reconciliation Commission (TRC), formally chaired by Archbishop Desmond Tutu, was a restorative justice commission established in 1995 to help heal the nation from the inhumane injustices of apartheid in South Africa.

The ideals of this commission may also apply to the establishment of a US Bioethics Commission on Truth and Reconciliation that could serve as a central body to document medical injustices and provide recommendations for reparations, healing, and truth-telling actions for victims and perpetrators of wrongdoing. For example, it recently was discovered that malnourished Aboriginals in Canada were uninformed and unprotected research subjects in nutritional studies in the 1940s and 1950s, during the same years in which the Tuskegee Syphilis Study took place. These government-supported experiments were "largely made possible because of access to a population of chronically malnourished and vulnerable children who, as wards of the state, had little say in whether or not they participated in the study" (Shuchman 2013, 1).

Ironically, these studies were partially funded by the Milbank Memorial Fund, which also financially supported the experiments in Tuskegee. As a result of these unethical experiments in Canada, the Truth and Reconciliation Commission of Canada won a case in January of 2013 that ordered the government to reveal documents about the studies (Shuchman 2013). Many of these records were destroyed, as in the case of the syphilis study in the United States (2013). But some documents are publicly available as archive and oral interview records (2013). As the commission continues its investigatory work, more details about these dental and nutritionally based studies on vulnerable and poor Aboriginal children will be revealed.

As in the case of Canada, the establishment of a US Bioethics Commission on Truth and Reconciliation can also help victims and their families in North America heal and reconcile past medical wrongdoings by revealing the intent, purpose, and goal of similar studies conducted and sponsored by the US federal government and other institutions. Furthermore, apologies and comprehensive reparations can be provided to victims because of unethical and unjust studies conducted on them. The commission would begin by first reviewing a more comprehensive compensation framework for Tuskegee victims and their families. The commission's duties would later extend to examining other unethical studies that may merit compensation and reparations for research subjects.

I propose the establishment of the US Bioethics Commission on Truth and Reconciliation because the current Presidential Commission for the Study of Bioethical Issues (the Bioethics Commission) and institutional review boards cannot provide distributive or restorative justice through reparation and compensation mechanisms for victims of unethical studies. Therefore, the new commission would serve as a practice-based entity that would be empowered to actually help heal victims from the outcome of unethical studies.

I argue that although victims of the experiments at Tuskegee were awarded monetary compensation, there are more long-term initiatives that could benefit future generations, such as scholarships and grants for family members of victims who want to receive a formal education. Education is one of the leading factors in the decrease of economic and health care disparities. Thus, a commission-based recommendation for compensation through educational opportunity can serve as a gateway to enhanced mobility in health, income equality, and higher learning attainment.

The African Concept of Ubuntu

Am I my brother's keeper? We are often asked this very question in our daily existence by how we treat one another, whether at school, in the workplace, or through our social engagements. In medicine and health care, when we conduct harm to a research subject in an experiment, we, essentially, harm ourselves as well as researchers and other human beings. More dramatically, when we allow unscientific categorizations, such as race, to bias our research outcomes, we, as scientists, pollute our research integrity. This basic notion can be summed up in the African concept of *Ubuntu*, which means "I am what I am because of who we all are" (Gbowee 2013).

Because the issue of race is so closely aligned to a people of African descent in this paper, it is most appropriate to discuss the African concept of Ubuntu. The concept is rooted in the belief that the acts I commit as an individual not only affect me but the whole or greater population. Therefore, if I harm another person or persons physically, psychologically, or spiritually, I, in turn, harm myself. This concept of Ubuntu was popularized, in large part, by the political philosophy of Nelson Mandela, who established the Truth and Reconciliation Commission in order to help bring reconciliation and healing to the nation of South Africa after the abolition of apartheid. President Mandela believed that in order to heal the nation from its human rights violations, forgiveness and reconciliation had to occur among all people affected by apartheid. Thus, the African concept of Ubuntu became a guiding principle in healing from the aftermath of apartheid. Mandela promoted the African concept of Ubuntu among everyday people and fought to bring South Africa together as a nation through forgiveness rather than revenge. He believed that revengeful acts for previous injustices would only further divide the country as a whole.

The African concept of Ubuntu can also be applied to the ethical realm of medicine, where unethical studies often ignore the value and human dignity of research subjects, which, in turn, also devalues the integrity and humanity of researchers. But fortunately, some medical organizations, institutions, and colleges are beginning to expose medical researchers, doctors, and other health care professionals to the importance of seeing the humanity of patients and research subjects. In turn, it is recommended that research entities integrate continuing education, learning workshops, and/or academic course requirements in the medical humanities in order to strengthen doctors' and researchers' knowledge of the medical value of patients and research subjects. In addition, teaching and implementing the African concept of Ubuntu in medical practice may also help increase empathy among researchers, serve as a model to systematically increase the enrollment of vulnerable populations in clinical trials, and help decrease health care disparities among racial minorities. It is also recommended that a comprehensive medical education curriculum guide be developed in order to teach emerging and seasoned researchers about medical and research practices that integrate narrative medicine and bioethics in research training.

An Apology

In the principle of compensatory justice, when an injustice has been committed, fair compensation or reparation is owned to the injured party (Beauchamp 1975). As Graham Hughes raises a similar view on reparations for blacks, the consideration of an apology after medical wrongdoings is also necessary to explore. Hughes frames it bluntly: "So if a public confession can be made in statutory form of the just claims of American Indians for compensation (a phenomenon uncomfortably close in time and nature to post-war Germany's reparations payments to Jews) why should we not initiate similar schemes for reparations to black Americans?" (Beauchamp 1975, 21). I go further to argue that an apology for medical wrongdoings must and should be inclusive of long-term compensatory outcomes rather than just monetary payments to victims.

The broader ethical issue is that of respect for persons. Did researchers genuinely respect vulnerable populations when they conducted their unethical medical deeds at Tuskegee? One cannot pay for respect, and it cannot be replaced or forgiven through monetary means alone. There has to exist more tangible and feasible means by which apologies for unethical medical actions are reconciled and compensated to victims. A broad means by which an apology can reconcile unethical practices committed against a vulnerable group is the establishment of a reparations system layered with policies that benefit as many members of that group as possible.

These policies can be inclusive of such programs as affirmative action programs that are geared toward rectifying past discrimination by initiating equal hiring practices. Some critics of affirmative action policies contend that they are forms of reverse discrimination and constitutionally illegal. I argue that the economic, political, and social system from which affirmative action policies spring forth is based on a racist, discriminatory caste system, which ultimately created the problem in the first place. The founding system was broken from the inception of America as a nation. So any means, for example, to correct medical wrongdoings is inherently imperfect, but that does not mean that the route by which to rectify the past is inherently unethical. It simply means that corrective measures, such as an apology for wrongdoings, are imperfect and subject to improvement according to current ethical and moral standards.

I became aware of the notion of an apology through a lecture I attended at Tuskegee University in April of 2011 featuring Carletta S. Tilousi. She presented a moving lecture about the Diabetes Project and how, similar to Tuskegee, this unethical experiment affected the Havasupai Native Americans in the Grand Canyon in Arizona. Tilousi is a member of the Havasupai Tribe and a participant in the Diabetes Project. In March 2003, she attended a lecture at Arizona State University where she learned that blood samples she provided to researchers were later used in studies without her consent or the consent of other tribal members (National Congress of American Indians 2013). During her presentation at Tuskegee, she shared with the audience that some of the repercussions for injustices that occurred to her people resulted in a reluctance of Indian tribes to participate in any human subject research project, Arizona State University Institutional Review Board (IRB) regulations being violated, and the loss of grant funding from such incurred violations. The overriding principle of justice was a theme that resonated throughout her lecture.

Although the Havasupai people have received compensation for what happened to them as a result of unethical human subject experimentation, Councilwoman Tilousi's presentation primarily highlighted that there still exists a lack of trust among Native Americans as a people to participate in any form of human subject research. When I met Councilwoman Tilousi after her lecture, I had the honor of speaking with her and taking a photograph with her and her colleague from Arizona (see appendix D). I talked with her about the similarities between the Diabetes Project and the Tuskegee Syphilis Study. I discovered that similar to Tuskegee University, Arizona State University has been unwilling to formally apologize for their role in sanctioning an unethical clinical trial.

In the case of Tuskegee, although an apology was provided by the federal government for its role in initiating the study, the actual medical actors involved have never apologized for their unethical medical practices. In fact, most medical doctors involved in the study have defended the study as an agent of scientific good rather than medical harm. Case in point, Dr. John Heller, director of venereal disease at the Public Health Service from 1943 to 1948, once stated: "I feel this was a perfectly straightforward study, perfectly ethical, with controls. Part of our mission as physicians is to find out what happens to individuals with disease and without disease" (Reverby 2000, 28).

Harriet Washington, author of *Medical Apartheid*, contends that our domestic wrongs have been perpetuated on black people without consequences. As she points out, the fear of involvement in medical studies by black people is rooted in real fears that have concrete merits. She adds that scientific racism contributed to the belief that unethical research conducted on black people was justified. Washington (2007) defines *scientific racism* as a whole set of beliefs by medical professionals that black people were medically different from other populations of people. It became a rational reasoning for slavery because medical doctors believed that blacks, for example, could work in difficult and harsh laborious conditions without fainting.

I argue that the apology given by the US federal government in 1997 through President William J. Clinton's address on medical wrongdoings at Tuskegee is insufficient in nature because the actual medical investigators involved in the experiments never apologized for their medical indiscretions. In fact, medical officials involved in the study did not lose their medical licenses or serve criminal sentences for their unethical medical practices. Some public health researchers even conducted similar unethical studies elsewhere during the Tuskegee Syphilis Study, such as unethical sexually transmitted disease studies conducted in Guatemala.

With all this in mind, how can a comprehensive apology for medical wrongdoings conducted at Tuskegee take place when the actual medical staff never apologized for their role in these unethical medical practices? Unfortunately, the study's medical architects are deceased and the known victims of the study are deceased as well. One of the worrisome myths of the study is that black medical officials involved in the experiment were victims of the experiment as well due to racism. I disagree. Black medical staff, such as Nurse Eunice Rivers and Dr. Eugene H. Dibble, Jr,. also played critical roles in deceiving the victims of the experiment. They were not the only black health officials involved in the study. But unfortunately, they simply represent how the involvement of black medical staff was widely promoted in order to increase the percentage of participants in the study. The fact remains that study participants trusted these black professionals because of their professional position in medicine. It is important to note that, historically, black medical professionals were revered by black patients because of the rarity of seeing such officials in black society.

As letter correspondences housed in at the Tuskegee University Archives Office and the National Archives Southeast Region in Marrow, Georgia, demonstrate, administration officials at Tuskegee University were well aware of the unethical merits of the Tuskegee Syphilis Study. From Robert R. Moton, Frederick D. Patterson, and Luther H. Foster, each university president of Tuskegee unilaterally agreed to allow the study to continue at the institute's hospital, John A. Andrew Memorial Hospital.

The recruitment grounds for the study were at black churches because the black church at the time was a sacred environment within the black community where trust was paramount. Nurse Rivers often recruited study participants for the syphilis study as evident in photographs of her and other white public health officials at recruitment church and school grounds, such as Shiloh Missionary Baptist Church in Notasulga, Alabama. These images are housed in the National Archives in Marrow, Georgia. I have depicted these images in my art collage series documenting the Tuskegee Syphilis Study. Author James H. Jones makes reference to this fact in his book, *Bad Blood: The Tuskegee Syphilis Experiment*. Black male study participants were also recruited from local communities, such as Hardaway and Shorter, in order to receive treatment for *bad blood* (Jones 1993).

In order to comprehensively understand black medical staff participation in the experiment, the documents confirming their involvement must be revealed. But unfortunately, many of these records that trace the involvement of black medical staff in the syphilis study have been destroyed. As Harriet Washington

(2007) revealed, the United States Public Health Service (USPHS) Study of Syphilis Ad Hoc Committee destroyed oral history testimonies collected from all parties involved in the experiment, including a testimony by Nurse Eunice Rivers, the nurse who served as a primary recruiter for study participants. The rationale for destroying the oral history tapes was that the committee did not want the tapes to implicate Nurse Rivers and bring her to trial (2007). The committee's unthinkable actions underlie the fact that full healing and closure through a comprehensive apology is difficult to achieve if medical officials could not even be held accountable for their implicit and unethical medical deeds, whether these officials were black or white. Institutions and even individual perpetrators involved in promoting ethical wrongdoings in the experiments at Tuskegee should critically analyze the issue and process of an apology that should be appropriately provided to victims and their family members who were affected by the negative outcomes of the study.

Clinical Empathy

Finally, challenges to justice in human subjects research can be decreased through the practice of clinical empathy. Some theorists view empathy generally as "recognizing and explicitly acknowledging the patient's emotion" (Cohen-Cole 1995). In the specific case of research, Halpern defines clinical empathy as emotional reasoning that is cognitive and emotional (Marcus 1999). Clinical empathy espouses that research practitioners be sympathetically immersed when making decisions about patients and must learn to use their emotional responses and their imaginations for therapeutic impact (1999). Simply stated, clinical empathy is the human quality of understanding the research subjects' health-related challenges or other factors, such as race, gender, and class, by symbolically taking their place in research experimentation.

In the case of human subjects research, too often, researchers do not view their research subjects as human beings with feelings, such as suffering and pain. Rather than only viewing research subjects as a means to an end or sample populations of convenience, I would recommend viewing research from the eyes of the subjects as well. Clinical empathy involves self-reflective questioning, such as: If I were a research subject, would I want to be informed that I had syphilis? If I were a research subject, would I want cell samples from my body extracted without my permission? If I were a research subject, would I want my blood to be used in another research project other than the one I was informed about? These are just a few of the questions we should be willing to ask ourselves before we, as researchers, conduct research on subjects.

If justice issues arise during a research experiment, it is recommended that, as a policy, researchers address these challenges by discussing them within the review of literature, recommendations, and/or further study sections of publishable research documents. In addition, researchers must be open to discussing justice challenges in research with colleagues in order to help institute principled research policy changes within research organizations. Therefore, clinical empathy is not only a reflective exercise, but it also serves as an active component for real and tangible change in research outcomes. As medical training specialists are beginning to understand, future medical researchers can be trained through humanities courses in medical school or continuing education to become more clinically empathetic in their research framework. Curriculum changes in medical training programs that emphasize clinical empathy will not only benefit medical researchers but enhance the research experience of participants in clinical studies.

Education

Another way in which justice challenges can be reduced is through education. Public health officials and bioethicists should work together to educate vulnerable research populations about unethical studies that the government conducted and discuss ethical procedures and guidelines currently available to protect such populations. Vulnerable populations will continue to be hesitant to participate in studies because of unfounded but, sometimes, justifiable misconceptions about research. For instance, although young black women and gay black men have been overrepresented among those living with and dying from AIDS (Fullilove 2006), they are less likely to participate in community-based research studies on HIV/AIDS due, in large part, to medical mistruth as a result of the unethical experiments held at Tuskegee and other historically unethical clinical trials. But it is also important to note that some research studies indicate the opposite effect. Therefore, the research in this area has produced mixed findings.

In order to decrease future misconceptions in research, it is recommended that public health officials and bioethicists develop an aggressive educational campaign to expose populations, such as gay black men, to the benefits and minimal risks of participating in human subjects research on HIV/AIDS. As a policy initiative, this campaign could be implemented by conducting educational workshops in worship centers, schools, and other institutions in order to recruit a diverse body of potential research study participants. In my own educational efforts, I have provided lectures and workshops to educational and religious institutions in Columbus, Georgia; Tuskegee, Alabama; and Harlem, New York, in order to educate people of color about the benefits of clinical research. In many instances, I teach audiences that the experiments at Tuskegee serve as a case study and an example of injustice in medical research. At the conclusion of my lectures, I often provide ethical and justice-based recommendations on how to reconcile past wrongdoings conducted in the Tuskegee Syphilis Study through future healing, reparations, and policy change.

The National Institute of Mental Health (2005) also provides several suggestions that address educational diversity issues related to minority population enrollment in mental health clinical research studies. These recommendations include that research staff should reflect the diversity of the groups desired for enrollment, establish relationships with respected members of the communities chosen for inclusion, work with a representative/liaison of specific communities to obtain ideas for enhancing communication, work with local churches and community centers, and offer all study materials in relevant languages (National Institute of Mental Health 2005). Ironically, the study held at Tuskegee implemented most of the recommendations even during the experiment's initial phases.

Other policy initiatives to help decrease injustice in research included reducing the erosion of informed consent in medical studies, especially in developing nations; increasing the effectiveness of institutional review boards; developing more comprehensive strategies to educate patients of informed consent, patient rights, and the importance of study participation, especially in communities involving vulnerable populations, such as minority groups; and allocating additional material resources in order to assist vulnerable populations that have been victimized in medical research (Washington 2007).

Additional general recommendations to help alleviate race and justice include the following:

1. Community engagement must also occur with research institutions and churches and other institutions in which vulnerable, ethnic, and minority populations reside and congregate. For example, health fairs, workshops, and other programs can help educate, inform, and reduce disparities in race and justice challenges and help increase the likelihood of human research participation in clinical trials.
2. Research investigators should seek to gain the trust of their potential research participants. Trust, in and of itself, is a

complex factor. In essence, informed consent forms are reflective of how one may trust researchers. Trust may be defined as a means by which good faith is evident. It is also the research subject's belief that he or she will not be taken advantage of or unjustly manipulated in medical or healthcare research.
3. One-on-one time with potential individual research subjects to help them better understand the details of the research study should be given. Such a process could enable the subject to more rationally provide informed consent.
4. The inclusion of minority educators in human research recruitment who can help interpret informed consent documents for ethnic minority and/or vulnerable populations will aid in increasing the possibility of these populations participating in human research experimentation. Minority educators can help visually and contextually translate relevant informed consent documents to potential minority and vulnerable populations who may decide to participate in human research experimentation.

Closing Statement

As we can see, race has affected every fabric of American life from law to clinical research. Although scientifically impossible to measure, race has been used as a weapon to prevent medical treatment for segmented and disenfranchised populations based on abstract, false, and unjustified medical beliefs in physical, mental, and spiritual inferiority. The first step in alleviating wrongdoings to any population is to recognize how harmful and even deadly race and injustice can hinder treatment for citizens who live among us in our society. The color of one's skin does not justify medical wrongdoings committed against another person. Likewise, the color of one's skin should never be a determining factor as to how treatment is provided or withheld. In the case of the Tuskegee Syphilis Study, race was the predominant factor in measuring whether black men should or should not receive treatment for syphilis. The sole selection of mostly black farmers to participate in the experiment was no accident. It was the greatest form of medical injustice. These men were victims of a social, economic, educational, and political system stacked against them simply because of the darkness of their skin.

As evident in my thesis, race and justice have been paramount factors in health as shown in the Tuskegee Syphilis Study. The study also highlights how even misconceptions, stereotypes, and racism within a minority population, such as the African American community, influence the perpetuation of unethical human subjects research. Race-based medical practices do little to solve the greatest disparities found in health and medicine. But bringing these barriers to light is a start to resolving challenges in race and health. Most of what we have learned about resolving issues surrounding race and health has been learned from mistakes of the past. It is my firm belief that the lessons of Tuskegee deserve to be learned and relearned in order to build a more sustainable and equal future for all our fellow citizens.

The Tuskegee Syphilis Study serves as a pertinent example of injustice in human subjects research. Although research participants were treated unethically and unfairly, the negative outcome of the study also brought forth the establishment of institutional review boards and informed consent policies that protect human research subjects even today. There are several lessons to be learned from the experiments at Tuskegee, such as how to ethically treat research subjects, how to conduct research based on the principle of justice, and how to use the mechanism of informed consent effectively to close injustice gaps in research, just to name a few.

For the arc of the moral universe to bend toward justice, as Dr. Martin Luther King, Jr., believed, we must be willing and able to clinically empathize with our most vulnerable citizens, while providing justice-based educational opportunities and resources to all people who participate in medical research. All in all, the Tuskegee Syphilis Study provides a remarkable case study of justice-related issues in medical and health care research. If understood properly and analyzed comprehensively, the experiments at Tuskegee serve as a means to rectify past medical wrongdoings through race and justice-based initiatives and solutions.

As in the voice of Nelson Mandela, we must see ourselves in others. We are called to discover the humanity in every individual. Liberian peace activist Leymah Gbowee once stated, "I am what I am because of who we all are" (2013). This concept of Ubuntu is ever so pertinent to the challenges we face to justice and equality in human subjects research. All humans are interconnected to each other. Therefore, our humanness must extend not only to everyday living, but to the very ideals of a fair and just research agenda.

References

American Medical Association. (2009). "Informed Consent." Retrieved from: URL: http://www.ama-assn.org/ama/pub/physician-resources/legal-topics/patient-physician-relationship-topics/informed-consent.Shtml.

"As Ye Sow, That Shall Ye Also Reap" (1899, June). *Atlanta Journal-Record of Medicine* 1, 266.

Beauchamp, T. L. (Ed.) (1975). *Ethics and Public Policy.* Prentice-Hall, Inc.

Belmont Report: Ethical Principles and Guidelines for the Protection of Human Subjects of Research (1979, April 18). The National Commission for the Protection of Human Subjects of Biomedical and Behavioral Research.

Brandt, A.M. (1978). "Racism and Research: The Case of the Tuskegee Syphilis Study." *The Hastings Center Report* 8(6): 21–29.

Carmichael, S. (1966, October). *Black Power.* Speech text. Retrieved from http://voicesofdemocracy.umd.edu/carmichael-black-power-speech-text/.

Centers for Disease Control and Prevention. (2014). "US. Public Health Service Syphilis Study at Tuskegee." Retrieved from http://www.cdc.gov/tuskegee/faq.htm.

Chang, H. (1974). "The Slave Economy in US Capitalism."

Chang, H. (1973). "Racism and Racial Categories."

Cohen-Cole, S.A. (1995). "Bird J. Function 2: Building Rapport and Responding to Patient's Emotions (Relationship Skills)," 21-7. In: Cohen-Cole SA, ed. *The Medical Interview: The Three-Function Approach*. St. Louis: Mosby Year Book.

Daniels, N. (2012). "Justice, Health, and Health Care." In *Medicine and Social Justice: Essays on the Distribution of Health Care*. 2nd Edition. Edited by Rhodes, R., M. Battin, and A. Silvers. Oxford University Press.

Daniels, N., B. P. Kennedy, and I. Kawachi. (1999). *Why Justice Is Good for Our Health: The Social Determinants of Health Inequalities*. The MIT Press.

"Deterioration of the American Negro." (1903, July). *Atlanta Journal-Record of Medicine* 5, 287–88.

Du Bois, W.E.B. (1900). "The Twelfth Census and the Negro Problems." *The Southern Workman* 29, 307.

Edgar, H. (1992). "Outside the Community." *Hastings Center Report*, 22, 32–35.

Frazier, E.F. (1933). "Children in Black and Mulatto Families." *American Journal of Sociology*, 39.

Frazier, E.F. (1957). *Black Bourgeoisie*.

Fullilove, R. (2006). "s, Health Disparities and HIV/AIDS: Recommendations for Confronting the Epidemic in Black America." *A Report from the National Minority AIDS Council*.

Fullilove, M. (1998, September). "Comment: Abandoning 'Race' as a Variable in Public Health Research—An Idea Whose Time Has Come." *American Journal of Public Health*, 88 no. 9, 1297–1298.

Gazzinelli, M.F., L. Lobato, L. Matoso, R. Avila, R. Marques, A. S. Brown, R. Correa-Oliveira, J. M. Bethony, and D. J. Diemert. (2010). "Health Education through Analogies: Preparation of a Community for Clinical Trials of a Vaccine against Hookworm in an Endemic Area of Brazil." *Neglected Tropical Diseases* 4 no. 7, 1–12.

Gbowee, L. (2013). Defining the Term, *Ubuntu*.

Graham, O.L. (1999). *Our Kind of People: Inside America's Black Upper Class*. Harper Collins Publishers.

Greene, S.M., and A. M. Geiger. (2006). "A Review Finds that Multicenter Studies Face Substantial Challenges but Strategies Exist to Achieve Institutional Review Board Approval." *Journal of Clinical Epidemiology* 59, 784–790.

Hare, K. (2010, January 28). "The Color Wheel: Rooted in History, Colorism Still Causes Prejudice Based on Skin Tone." *The St. Louis Beacon* (online journal).

Hazen, H.H. (1914, August, 8). "Syphilis in the American Negro." *Journal of the American Medical Association* 63, 463.

Hochschild, J.L., and B. M. Powell. (2008). "Racial Reorganization and the United States Census 1850–1930: Mulattoes,

Half-Breeds, Mixed Parentage, Hindoos, and the Mexican race." *Studies in American Political Development* 22 no. 1, 59–96.

Hochhauser, M. (2004, April). "Informed Consent: Reading and Understanding Are Not the Same." *Applied Clinical Trials Magazine*, 221–225.

Jones, J. H. (1993). *Bad Blood: The Tuskegee Syphilis Experiment.* The Free Press.

Klitzman, R. (2013). Lecture on the Role of IRBs. Research Ethics Course at Columbia University, New York, New York.

Krieger, N. (2003, February). "Does Racism Harm Health? Did Child Abuse Exist before 1962? On Explicit Questions, Critical Science, and Current Controversies: An Ecosocial Perspective." *American Journal of Public Health* 93 no. 2, 194–199.

Levine, R.J. (1984, September/October). Case Study: "What Kinds of Subjects Can Understand This Protocol?"

Locke, J. (1690). *Two Treatises on Government.* The Project Gutenberg E-book.

Mackintosh, D.R., and V. J. Molloy. (2003, May). "Opportunities to Improve Informed Consent: Frequently Observed Problems in Processes and Consent." *Applied Clinical Trials*, 42–48.

Marcus, E.R. (1999). "Empathy, Humanism, and the Professionalization Process of Medical Education." *Academic Medicine* 74:1211–5.

National Bureau of Economic Research. (2005). "Colorism and Wealth" [conference paper].

National Congress of American Indians. (2013). "Havasupai Tribe and the Lawsuit Settlement Aftermath." *American Indian and Alaska Native Genetics Resource Center.*

National Institute of Mental Health (2005, June). "Points to Consider about Recruitment and Retention while Preparing a Clinical Research Study."

NA-WRC. (1933, July 27). *Cumming to Moton* (letter correspondences).

NA-WNRC (1933, August 5). *Wenger to Vonderlehr* (letter correspondences).

Nobles, M. (2000). *Shades of Citizenship: Race and the Census in Modern Politics.* Stanford University Press: Stanford, California.

Pollock, J.M. (2005). "The Rationale for Imprisonment." *The Philosophy and History of Prisons Handbook.*

Prime Time Live (1992). "The Tuskegee Study" [video]. ABC News.

Reverby, S. (2000). *Tuskegee's Truths: Rethinking the Tuskegee Syphilis Study.* Chapel Hill and London: University of North Carolina Press.

Russell, K.Y., M. Wilson, and R. E. Hall. (1992). *The Color Complex: The Politics of Skin Color among s.* First Anchor Book Edition.

Sarkar, R., T. V. Sowmyanarayanan, P. Samuel, S. S. Singh, A. Bose, J. Muliyil, and G. Kang. (2010). "Comparison of Group Counseling with Individual Counseling in the Comprehension of Informed Consent: A Randomized Controlled Trial." *BMC Medical Ethics*, 11 (8). Retrieved from http://www.biomedcentral.com/1472-6939/11/8.

Schenker, Y., A. Fernandez, R. Sudore, and D. Schillinger. (2011, January/February).

"Interventions to Improve Patient Comprehension in Informed Consent for Medical and Surgical Procedures: A Systematic Review." *Medical Decision Making*, 151–173.

Shuchman, M. (2013, December 9). "Canada Confronts Its Own 'Tuskegee' Studies." *Bioethics Forum*: The Hastings Center, 1–2.

Thomas, S.B., and S. C. Quinn. (2000, July). "Light on the Shadow of the Syphilis Study at Tuskegee." *Health Promotion Practice* 1 no. 3, 234–237. Sage Publications.

Washington, B.T. (1901). *Up from Slavery*. Doubleday Publishing Company.

Washington, H. (2007, February 7). Lecture on *Medical Apartheid: The Dark History of Medical Experimentation on Black America*. Enoch Pratt Free Library: Baltimore, Maryland. Book TV C-Span 2.

Wing, B. (2007). "Harry Chang: A Seminal Theorist of Racial Justice." *Monthly Review*. Retrieved from http://monthlyreview.org/2007/01/01/harry-chang-a-seminal-theorist-of-racial-justice.

X, Malcolm. (1963, January 23). *The Race Problem: Malcolm Describes the Difference between the "House Negro" and the "Field Negro."* Audio recording: Michigan State University, East Lansing, Michigan.

Appendices
Applied Experiences as an IRB Graduate Intern

During my internal review board (IRB) internship, I was provided a plethora of experiences that were both practical and theoretical. It was an honor to be selected in the first cohort of bioethics graduate students at Columbia University to participate in the internship program. My internship began in the fall 2013 semester. I chose to conduct my internship at the New York State Psychiatric Institute. Within the first weeks of my internship, I observed several IRB committee and subcommittee meetings at the institute. During these meetings, I learned how and why psychiatric and psychological clinical trials were approved, returned back to research investigators for revisions, or not approved at all. In summation, many studies were approved or not approved based on the risks and benefits ratio to research subjects. I learned that the benefits, when possible, should always outweigh the risks involved in a study. When possible, risks to the research subject should be reduced. At times, the patient may not receive any personal or tangible benefits. In the long term, benefits may also include the subject's contribution to scientific discovery, which in turn, can serve as a benefit to society as a whole.

As I learned more about the benefits and risks ratio, I began to clearly understand how the Tuskegee Syphilis Study violated this fundamental ethic. Clearly, the research participants in the experiment did not receive a valuable benefit since they were not treated for syphilis or informed properly that they had the disease. As an emerging bioethicist, I now clearly understand how this vital concept of risks versus benefits connects to the atrocities that occurred in experiments at Tuskegee.

Another area during my internship that I discovered was significant to the IRB approval process was informed consent. While attending some IRB subcommittee meetings, I observed that some research protocols were sent back to investigators because the informed consent section was inconsistent, unclear, or excluded pertinent details that potential research subjects needed to be aware of in order to decide if they wanted to participate in the study. In some cases, minor inconsistencies were the biggest errors. As I attended these IRB meetings, I began to realize how vital IRBs are in identifying errors, inconsistencies, or the exclusion of important details in informed consent documents. In addition, I also realized that informed consent forms should be clear enough in order for potential research subjects to make reasonable decisions as to whether or not they should participate in a research study. The most egregious and unethical act that affected the research subjects' decision to participate in the Tuskegee Syphilis Study was the fact that these men were not properly and clearly provided informed consent about the true nature of the study and their role as participants. Furthermore, research subjects who had syphilis were not properly informed that they had the disease. In addition, they were not informed about treatments for syphilis even when penicillin was available to help syphilitic subjects. In essence, I discovered that this type of medical deception serves as another form of injustice in health care research because the most vulnerable are purposefully misguided into making irrational and uninformed health care decisions without proper consultation.

During my internship, I was able to accomplish a lifelong goal of publishing a book with my brother, Dr. Ejinkonye C. Anekwe, about the Tuskegee Syphilis Study. The book entitled, *Chronicling the Tuskegee Syphilis Study: Essays, Research Writings, Commentaries, and Other Documented Works*, was independently published in December of 2013 (see appendix E). The book is a compilation of writings we both wrote over a period of about five years. I was greatly pleased that the publication of our book occurred during my internship experience because I was able to use the book as a resource guide for my thesis. The process of completing the book also provided closure for me personally on a subject matter in which I was personally invested as a person born in the very hospital where the Tuskegee Syphilis Study was conducted. During the tenure of my parents working at Tuskegee Institute (now Tuskegee University), I was born at the John A. Andrew Memorial Hospital on the campus in 1974. My father, Dr. Gregory E. Anekwe, was a biochemistry researcher and associate professor while my mother, Ms. Emma J. Anekwe, worked as a medical technologist in the hospital where I was born. Because of my personal connection to the location in which the experiments at Tuskegee were conducted, I felt I had a personal responsibility to educate the general public about the race-based injustices that occurred in the study. In essence, our book became a vehicle to formally highlight and explore race and justice challenges in clinical research through an analysis of the Tuskegee Syphilis Study. The book became a voice for the voiceless and a vehicle to transform a human atrocity into a means by which genuine healing can occur.

Another way in which I have been able to educate the public about race and justice challenges within the Tuskegee Syphilis Study has been through my artwork. The cover for our book was a collage entitled, "Human Subjects," which is part of my thirty-piece art collage series about the Tuskegee Syphilis Study. The art series serves as a tool to connect with people who learn best through visual interpretations. I have also been able to reach a more diverse audience to inform them about race and justice challenges in medical research through publications such as *Academic Medicine*. This journal published my art collage entitled, "Doctor-Patient," on the cover of the December 2013 issue with my art summary about the interpretative nature of my artwork (see appendix F).

My goal has been to educate African Americans about their human rights as potential research participants and the various measures that are in place, such as informed consent, to protect their rights as human subject participants. Another reason why I have focused on presenting my research on the syphilis study to the black church is because many participants in the study were actually recruited from the church. Therefore, many African Americans still have reservations about participating in public health clinical trials because of unethical studies, such as Tuskegee. I have strived to inform the African American community in the black church that participating in clinical research is important and beneficial because a significant number of men and women in the black community have been disproportionately infected with HIV/AIDS and various forms of cancer. The lack of participation in clinical trials is detrimental to our health as a black community. As a result, many African Americans have been infected or even died from diseases for which treatment could have been available to medically help them if they were aware of these choices.

I have often asked myself what role I could play as an emerging bioethicist in helping decrease disparities concerning race and justice challenges in medical research. Some of my own personal initiatives and solutions have already been discussed. But I would like to add that my educational journey has taken me to the black church. Since 2013, I have spoken at various African American churches in Brooklyn and Harlem, New York, about the Tuskegee Syphilis Study. Bethany Baptist Church, in particular, has invited me twice to speak about the Tuskegee Syphilis Study. Most recently, I spoke at their Black History Month program on February 16, 2014. My church presentation on this date also served as a discussion ground for my book since its publication (see appendix G). My acts of community engagement have enabled me to connect with an audience of people who otherwise would not have known about their rights as potential research subjects.

Thus, I developed the Black Church Healing Model (see display 3) as a viable means by which race-based justice can be increased in the black community through working with the black church. In many respects, the black church represents the spiritual, educational, and social epicenter of the black community. Through the black church, vulnerable populations can be effectively informed about the benefits of preventive health care practices and participation in clinical trials. The model focuses on the formation and implementation of healing circle discussion groups and health care workshops / health fairs as viable mechanisms to enhance restorative and race-based justice. According to my model, recommended solutions and implementation strategies based on agreed-upon issues of discussion from church members also systematically increases restorative and race-based justice.

Display 3

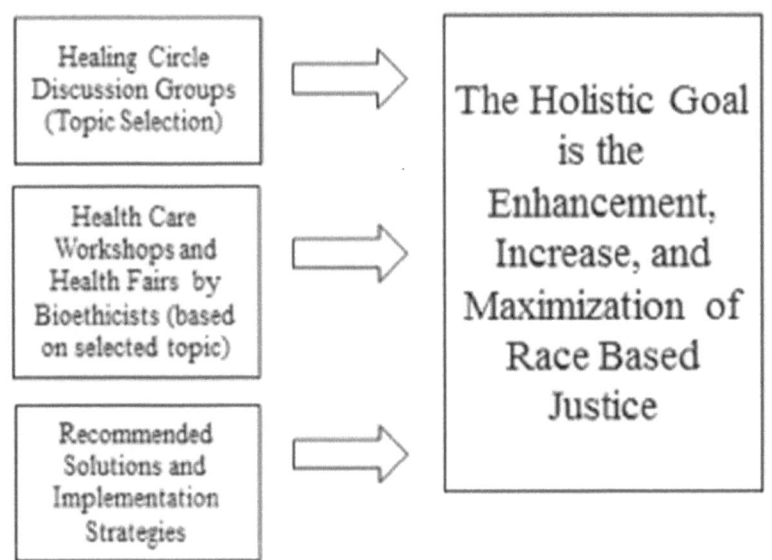

Additional Documents Attached

Appendix A: Industrialist Andrew Carnegie (front row, center) financially supported the Tuskegee Institute and its faculty members, pictured here. Carnegie lauded the efforts of Booker T. Washington, who opened the school in 1881, shown here with his wife, Margaret, next to the businessman. (Library of Congress Prints and Photographs Division)

ISSUES OF RACE, JUSTICE, GLOBAL COLORISM, SKIN BLEACHING, AND INTERNALIZED PERCEPTIONS OF SKIN COLOR

Appendix B: Dr. Eugene H. Dibble, Jr., the chief administrator of John A. Andrew Memorial Hospital, sanctioned the Tuskegee Syphilis Study to be conducted at the hospital that stood on the campus of Tuskegee Institute (now Tuskegee University). At the time of the syphilis study, Dr. Dibble was a mulatto whose family was one of the most prominent black families in the nation. Photograph courtesy of the private library of Louis Rabb.

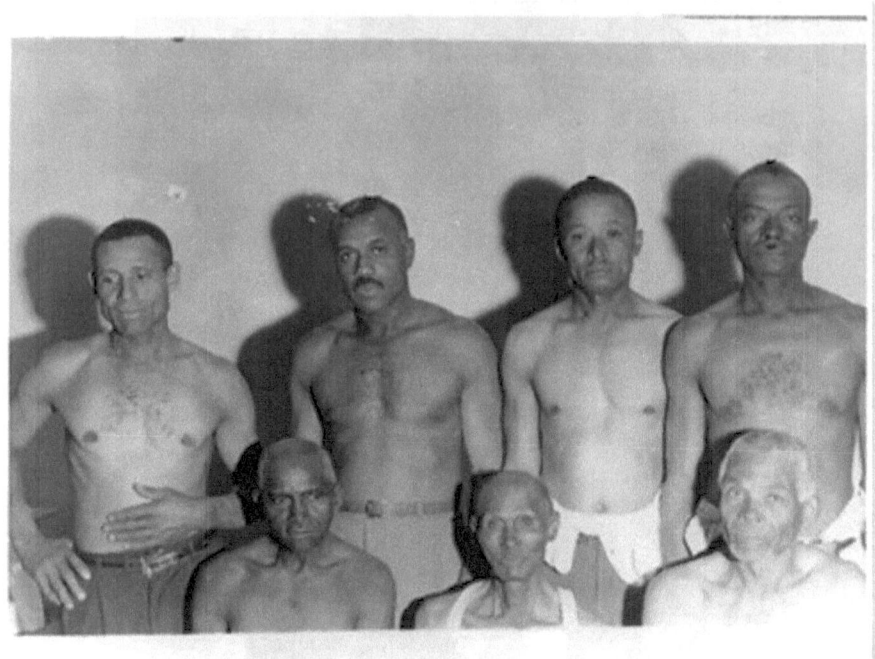

Appendix C: Many men subjected to the Tuskegee Syphilis Study were from the Southern farming class of s directly descended from the field Negro population of enslaved people in the South. Photograph courtesy of National Archives and Records Administration, Southeast Region, Morrow, Georgia.

ISSUES OF RACE, JUSTICE, GLOBAL COLORISM, SKIN BLEACHING, AND INTERNALIZED PERCEPTIONS OF SKIN COLOR

Appendix D: Pictured from left to right—Havasupai American Indian dressed in native attire, Obiora N. Anekwe, and Carletta S. Tilousi, Havasupai Tribal Council member at the Tuskegee Syphilis Study Apology Commemoration Week held on April 16, 2011, at the Kellogg Conference Center, Tuskegee University.

Appendix E: Image of book cover entitled, *Chronicling the Tuskegee Syphilis Study: Essays, Research Writings, Commentaries, and Other Documented Works*. The book was published in December 2013 during my graduate studies in bioethics at Columbia University.

ISSUES OF RACE, JUSTICE, GLOBAL COLORISM, SKIN BLEACHING, AND INTERNALIZED PERCEPTIONS OF SKIN COLOR

Appendix F: Art collage entitled, "Doctor-Patient," from my art collage series on the Tuskegee Syphilis Study. The artwork was featured on the cover of the December 2013 issue of *Academic Medicine* during my graduate studies in bioethics at Columbia University.

Appendix G: Pictured at Bethany Baptist Church in Harlem, New York, on February 16, 2014, after a program presentation for Black History Month during church service. During my presentation, I discussed my new book shown here and spirituality within the Tuskegee Syphilis Study and the black church.

ISSUES OF RACE, JUSTICE, GLOBAL COLORISM, SKIN BLEACHING, AND INTERNALIZED PERCEPTIONS OF SKIN COLOR

Global Colorism: An Ethical Issue and Challenge in Bioethics
Obiora N. Anekwe, Ed.D, M.S (Bioethics)

Introduction

Global colorism is seldom discussed in the field of bioethics, but it affects almost every facet of medical practice. The ethical challenge of colorism has global implications that are psychologically, physiologically, sociologically, and medically related. One impacts the other, sometimes without much notice. My goal for this paper is to bring issues related to colorism to light in order to begin the process of holistic healing for people of color, who are often the most vulnerable in health care and medicine. Secondly, I hope that this discussion of global colorism will bring forth a greater realization that more research needs to be conducted on the various impacts of colorism in medical practice.

What Is Colorism?

According to Baruti (2000), colorism is a global prejudice that people of African ancestry have toward each other and seemingly use against or to the advantage of themselves and others of relatively similar complexion. Herring (2004) also defines colorism as "discriminatory treatment of individuals falling within the same 'racial' group on the basis of skin color" (21). In order to provide a more expansive definition of global colorism, I would like to emphasize in this paper that global colorism may also be promoted and practiced by the oppressed (in this case, people of color) and the oppressor (postcolonialist Caucasians). As such, the term *racism*, for the purposes of this paper, is referred to as discriminatory power dynamics externally expressed by postcolonialists of European descent through decision-making practices based on the skin tone and complexion of oppressed people of color.

The History of Skin Color

The various skin colors of humans that currently exist evolved over the past sixty thousand years as humans dispersed out of equatorial Africa and adapted to new environments (Jablonski 2012). Specifically, scientific records confirm that all humanity originates from the Khuiland Great Lakes region in northeast Africa before seven million years ago (Bynum, Brown, King, and Moore 2005). Subsequent migrations occurred east to Ethiopia, north to Egypt and Eurasia, south to South Africa, and west to Chad, West Africa, and North Africa (2005). Thousands of years later, humans migrated out of Africa onto every continent on the planet (2005).

Because ecosystems in Africa changed beginning 2 to 1.5 million years ago, weather conditions became highly seasonal. These unpredictable changes produced transitions from forest to woodland and woodland to grassland (Jablonski 2012). As a result, plants and animals either evolved to cope with these changes in their environments or they became extinct (Jablonski 2012). As the late archaeologist J. Desmond Clark cited, human populations followed the seasonal migrations of the herds they depended on (2012). This action drew some populations of people northward and eastward out of Africa into Asia. By 1.8 million years ago, these people were living in central and eastern Asia (2012). As indicated in image 1, this Facebook advanced image shows the interconnection of the human race on earth throughout thousands of evolutionary years. Note that the divided regions of the earth can fix approximately as a whole like a puzzle if these regions are digitally put back together. This phenomenon can be explained because the earth once was one collective body of land that eventually divided into various regions or continents because of environmental climate changes.

ISSUES OF RACE, JUSTICE, GLOBAL COLORISM, SKIN BLEACHING, AND INTERNALIZED PERCEPTIONS OF SKIN COLOR

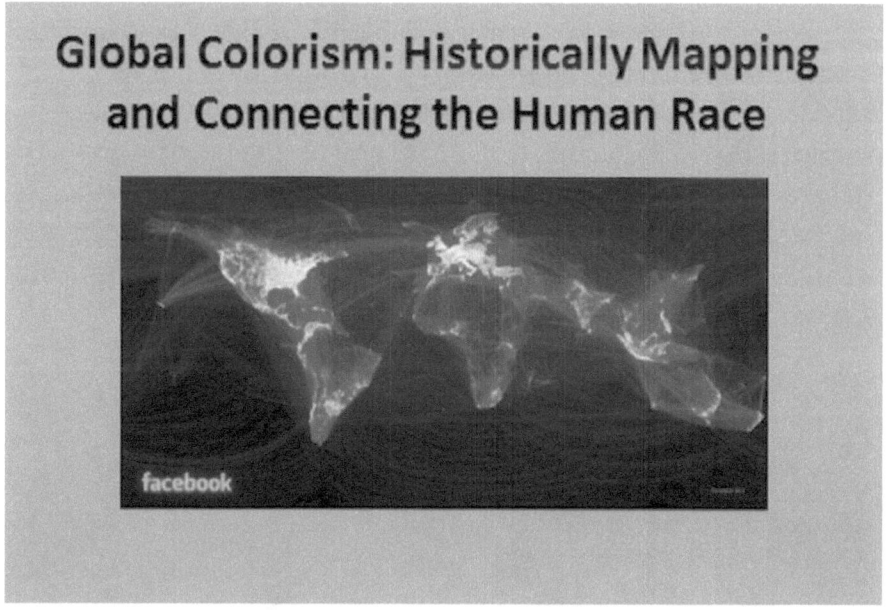

Image 1: Facebook advanced image of the earth's interconnection of the human race.

Thus, it is reasonable to conclude that skin color in humans evolved over many centuries to suit the environments where people eventually settled (Jablonski 2012). As Jablonski (2012) notes:

> The first hominid dispersal out of the tropics of Africa required changes in bodies, behavior, and culture to adapt to new environments and respond to new threats. The intensity and yearly pattern of sunshine in tropical Africa is, and was, very different from that of most of Asia and Europe.

Research estimates that it took ten to twenty thousand years for human populations to achieve the optimal level of pigmentation for the regional areas they settled (Jablonski 2012). It is significant to note that people living in Africa have high levels of mixed ancestry and display more genetic diversity than all the rest of the people in the world combined. This type of diversity is most visible through skin pigmentation (see images 2 and 3). Nevertheless, this variation in skin pigmentation within Africa is reflective of natural selection, migration, and the creation of new pigmentation mutations (2012).

Image 2: Collage of West African man who is highly melanated in skin complexion.
Collage: Obiora N. Anekwe

ISSUES OF RACE, JUSTICE, GLOBAL COLORISM, SKIN BLEACHING, AND INTERNALIZED PERCEPTIONS OF SKIN COLOR

Image 3: Collage image of my Nigerian aunt, Dr. Ngozi, MD, a medical doctor who resides in London, England. Her skin complexion is olive brown.
Collage: Obiora N. Anekwe

The Origins of Colorism

Colorism mostly originates from the mental and physical enslavement of black people in the Western Hemisphere. For example, in 1712, Willie Lynch gave a speech in Virginia to teach his proven and effective methods of enslaving black people in the Americas. He was the British slave owner in the West Indies from whom the term *lynching* is derived to indicate when people were hanged on trees by a mob of people.

In *Willie Lynch Letters: The Making of a Slave,* Lynch (1712) revealed that segregating the black race according to color served as the most effective means by which slave owners could control and manipulate their black slaves. In order to become an effective slave owner, Willie Lynch (1712) argued that "you must use the dark skin slaves vs. the light skin slaves, and the light skin slaves vs. the dark skin slaves…They must love, respect and trust only us…the slaves themselves will remain perpetually distrustful of each other" (1).

Early in the colonization of North America, slave owners began to do as Lynch directed. They separated the light-skinned, mixed-race (mulatto) slaves and housed them in the master's house. These slaves worked in the home of their slave master because, oftentimes, they were the biracial children of their slave master (Ulen 2013). On the other hand, darker-skinned black slaves worked in the outdoor fields and lived in the external living quarters (shanties) near the big house or slave master's home (Ulen 2013). This system of color segregation helped establish and promote what we now know as colorism within the international black community. As Willie Lynch's message of dividing and conquering black enslaved people spread throughout North America, it also gained acceptance as an effective means to guarantee the enslavement of other colonized people around the world.

Colorism and racialized discrimination of African people were two dominant factors involved in justifying institutionalized slavery in the Americas. It essentially was a savvy political move to maintain the transatlantic slave trade (Jablonski 2012). In the American War of Independence, the enslavement of blacks was one of the most significant factors in fighting for and defending American freedom (2012). Needless to say, there is an inextricable relationship between slavery, slaveholding, and the founding of the United States of America (2012).

The strategically planned dehumanization of Africans began when colonialists first associated blackness as evil and negative through a series of brutal and negative images and mythical language. For example, colonialists often used scripture in the Bible as a justification of slavery. In the account of Cain's exile, scripture was inaccurately interpreted to infer that the black descendants of Cain were cursed, when, in fact, the mark of Cain conferred God's protection and involved a curse on any person who harmed Cain in his exile (Jablonski 2012). It was only later when Christian colonialists purposefully misinterpreted the mark itself as a sign of punishment, shame, and corrupt meaning that black people were a cursed race (2012).

Race Constructs: Sociological, rather than Scientific

In a lecture at Harvard University, Nell Irvin Painter (2012), author of *The History of White People* (2010), pointed out that "race is an idea, not a fact." Trautmann (2004) rightly asserts that "race is socially constructed, that it is not objective but conventional, and that, therefore, it has contingent, historical character that is not perduring but governed by force in play at a given time" (214). Therefore, there is no scientific evidence to elude to the belief that race is a biological quality. Race is an abstract and flexible notion. It is sociologically, rather than biologically, based. The meaning of race has also translated into interpreting physical characteristics (Painter 2012). We mostly view beauty, especially light skin complexion, as personal characteristics we deem as attractive and worthy of deserved attention (2012).

The age-old question of whether race determines genetic differences has been debated for decades. In 2000, the Human Genome Project concluded that there is virtually no genetic difference between people of different skin colors (Rogers 2010). In fact, scientists have discovered that all humans share 99.99 percent of the same genetic code, no matter the race of the person (Rogers 2010). Geneticist J. Craig Venter concluded that this fact proves that race is definitely a "social concept, not a scientific one" (2010, 1). Nell Irvin Painter notes that the concept of whiteness originates from ancient Rome, where slaves were ironically, mostly white. The elevation of some ethnic populations, such as Germans as "whiter" than other groups originates from scientists who measured human skulls and set abstract criteria for beauty in people (2010).

Colorism around the World

Brown-Glaude (2007) notes that in European-colonized countries, like Jamaica and Tanzania, there are perceptions that light-skinned people in the middle and upper classes typically fair better during tough economic times than do darker-skinned people. In fact, these groups have even thrived during economic hardships (Lewis, Harris, Camp, Kalala, Jones, Ellick, Huff, and Younge 2013). Therefore, this occurrence actually translates into real economic, social, and health care advantages.

In Indian society, for instance, colorism affects just about every social, political, and health-related stratosphere. The caste system in India is a unique form of social division in society. For instance, the Aryan race (composed mostly of the upper class in India) was historically largely equated with the European/Caucasian race. They were considered as warriors who were physically sturdy and handsome because of their lighter skin, blue eyes, and sharp noses. The Aryans were mostly the European invaders of India whose descendants now occupy the upper class. In contrast, the indigenous darker-skinned people of India were described as ugly because of their dark skin, as extensively documented in the canonical text of the Hindus. Many of these religious writings expound on the racial distinction between lighter-skinned Aryans, who migrated into India from the north, and dark-skinned indigenous Indians of the south (Glenn 2009). Thapar (2008) argues that the term *Aryan* refers to a specific term of language and a sign of social superiority that has been misused to refer to race. These views have been evident in reinforced ideals of social Darwinism and biological determinism.

One of the negative side effects of colonialism in India and the Americas was a divide-and-conquer social order based on such external attributes as skin color, hair texture, and facial structure (Khan 2009). This well-devised form of racial categorization was established in order to maintain social order. In Trinidad, for instance, race-based, unscientific social orders were promoted in order to maintain social, legal, and moral order (2009). Most Caribbean nations, such as Trinidad, held a nationalistic and racial scientific belief that race was both biologically and socially heritable (2009). Although these belief systems still exist today, some scholars contend that such localized coloristic stratifications existed well before European colonialism (2009). As John Rogers points out, "ideas of difference with a quasi-biological character were already prevalent before the beginning of British rule, and many of the symbols and labels propagated in the name of racial ideology were drawn from earlier periods" (Khan 2009, 103). It is interesting to note that after colonialism, there has been an increased presence in Afro-Trinidadian traditional heritage, such as calypso and steel-pan music, the Orisha religion, and Carnival celebrations (see images 4 and 5). As one can observe in images 4 and 5, celebrations, such as Carnival, highlight the diversity of skin colors and the influence of the Yoruba religion on Caribbean culture.

Image 4
Collage: Obiora N. Anekwe

Image 5
Collage: Obiora N. Anekwe

Skin-Tone Bias and Colorism in Egg Donations for Assisted Reproduction

The practice of skin-tone bias and colorism has impacted the practice of donating eggs for assisted reproduction. In an effort to preserve skin tone, some egg-donor recipients base their selection on the skin color of egg donors. Skin tone is one of the many categories by which some egg donor databases characterize and differentiate donors (Thompson 2009). For instance, a source of skin tone signaling in structuring commercial US egg donation is through donor photographs and biographical detail (2009). In most cases involving sperm donation, photographs of a donor as a child are common visual representations (2009). In contrast, adult rather than baby photographs of egg donors are often provided to potential donor recipients.

These examples highlight how potential donor recipients are able to unscientifically determine the potential skin color of their future child. This practice, of course, is not guaranteed to produce babies with lighter skin, but it does establish a framework through which donor recipients may produce lighter-skinned babies. Therefore, a consideration of skin tone in egg donation illustrates the process of racializing biology for the sake of light-skinned preservation (Thompson 2009). Further research in bioethics needs to be conducted in order to explore this vital and significant issue in global and medical colorism.

The Physical, Sociological, and Psychological Impact of Colorism

Psychiatry serves as an ample foundation on which to discuss global colorism because what we think about ourselves often affects what we eventually do. A critique of mental behavior can ultimately help explain the practice of internalized skin discrimination within certain groups of colored people. According to psychiatrist Frantz Fanon (1952), darker-skinned people who have been oppressed as a result of colonialism tend to later take on the psychological, spiritual, and physical attributes of their oppressors. For instance, the practice of lightening one's skin is a physical expression by the oppressed (black people) to assimilate and become as their oppressors (Fanon 1952). They find little to no value in their own cultural expressions of selfhood but rather depend on the spiritual, physical, and psychological characteristics (language, religion, dress, etc.) of their oppressors in order to define and validate their human existence (1952). In fact, one can argue that the reason why people of color were and continue to be oppressed by the dominant power structure is because of the collective human ego's desire to dominate and control the most vulnerable by any means necessary, even to the extent of promoting nonscientific beliefs about inferiority based on skin color.

Noted psychiatrists, such as Frantz Fanon and Frances Cress Welsing, explain that colorism in medical practice is inherently based on self-hatred and low self-esteem rooted in colonialism and white supremacy. In many ways, the practice of colorism is a form of self-hatred that ultimately causes great harm to the oppressed, not the oppressor. For instance, rather than ridding oneself of his or her dark, melanated skin through the act of skin bleaching, greater physical and psychological harm is caused to those who are the most vulnerable in health care. But, on the other hand, one can also argue that because of the salient effects of global colorism, it impacts all human interactions through medical challenges, such as negative mental health and psychological implications of low self-esteem found among some dark-skinned women.

Colorism systematically seems to affect many areas of medicine and health care practice. For instance, if a person with highly melanated skin has internalized a belief system that says that he or she is inferior based on the darkness of skin color, then he or she is more likely to rid him- or herself of this physical attribute by eliminating it in order to psychologically feel better about his or her own human existence. Unfortunately, colorism even affects how research is framed, what is studied, who is studied, who is selected as research subjects, and how they are selected. Medical historian Harriet A. Washington (2007) most vividly documents this belief in her book *Medical Apartheid*, which is considered one of the most comprehensive historical critiques about unethical medical practices conducted on black people in America.

ISSUES OF RACE, JUSTICE, GLOBAL COLORISM, SKIN BLEACHING, AND INTERNALIZED PERCEPTIONS OF SKIN COLOR

Harvard-trained sociologist W. E. B. DuBois was one of the first scholars to address the psychological conflict that people of color have to encounter in their attempts to assimilate to the dominant white social group (Myrdal 1944). He often referred to this particular phenomenon as a "double consciousness," indicating that it was a response to cultural domination in a color-ranking society (Lewis et al. 2013). DuBois (1903) noted in his breakthrough book, *The Souls of Black Folk*, that "the problem of the twentieth century is the problem of the color-line—the relation of the darker to the lighter races of men in Asia and Africa, in America and the islands of the sea" (1). This problem has persisted even throughout the twenty-first century.

Unbeknownst to many, colorism has unfortunately been a determining factor in unethical, discriminatory practices in scientific inquiry. It has negatively affected people of color and affected the way in which science is practiced. But most significantly, it has affected the perspective from which darker-skinned people have viewed themselves, which is often manifested in internalized communal practices, such as colorism within the black and brown communities of color throughout the world.

Psychiatrist Frances Cress Welsing's theory of color confrontation helps to explain the negative practices of colorism and the strategic suppression of highly melanated people around the world. Her theory is based on the research analysis of race theorist Neely Fuller, Jr., who argued that racism is a system woven into the very fabric of American society (Welsing 2013). Fuller concluded that racism and white supremacy are articulated and practiced in every social system (Welsing 2013), including for our discussion, science and medicine. Welsing (2013) argues that white supremacy and racism are systems built on the global view of the white minority class that highly melanated people around the globe must be suppressed and even eliminated in order to maintain white domination. In essence, we can see how such a system is actualized through decades-old race-based medical practices.

Racism, as Welsing (2013) contends, is woven into the very fabric of human society, especially within American society, which was distinctly built upon the physical and now mental subjugation of people of a darker hue. She concludes her analysis by arguing that black suppression by the white race is a practiced form of psychological projection of white fear and inferiority (Welsing 2013). In other words, the white population's internalized inferiority complex and global fear of population extinction has been historically masked through vicious and violent acts of racism, white supremacy, human enslavement, and territorial colonization.

Historically, a miseducation about skin color has persisted over the centuries. In fact, the enslavement of dark-skinned people was systematically justified because slave owners promoted the nonscientific concept that blacks were genetically inferior and less human than white people. Although geneticists have consistently demonstrated that race is biologically unsubstantiated, this belief unfortunately still exists among some scholars (Funderburg 2013). In fact, it was German scientist Johann Friedrich Blumenbach who established racial categories according to skin color in the late eighteenth century (2013). This form of color classification was eventually used as a mechanism to divide and conquer the hearts and minds of melanated people in order to continue the slave trade and justify the suppression of the darker race throughout the world.

Studies Related to Colorism

Skin color affects life outcomes in physical health, mental health, and interpersonal relationships (Luisa et al. 2006, Eric et al. 2002, and Wade and Sara 2005). Wilder and Cain (2011) argue that colorism is perpetuated more frequently by women. It has caused a significant negative psychological impact during the childhood and adolescence of females (Wilder and Cain 2011). Stephens and Few (2007) have also examined skin color in relation to adolescent perceptions of sexual attractiveness in women via music videos and found that African American boys believed that light skin is the most significant factor in beauty ideal. De Casanova (2004) conducted a colorism and self-esteem study on Ecuadorian Latina adolescent girls and discovered that Ecuadorian girls describe the beauty ideal as "tall, thin, long yellow hair, and light eyes" (296).

Colorism seems to affect every aspect of our social existence, from employment hiring practices to prison sentencing. For example, in the United States, South Africa, the Caribbean, India, the Philippines, Korea, Latin America, and Mexico, skin color is a major factor in determining one's social stratification and its outcomes, whether in education, employment opportunities, income, or health status (Jablonski 2012). It has had a negative psychological and physical impact on mainstream America's perceptions of personhood to such an extent that skin color is more significant than more tangible human qualities. For example, a Johns Hopkins University study in 2011 of US Census data concluded that mixed-race people are socially placed below whites but ahead of blacks (Smith 2011).

Moreover, a recent Villanova University study in 2011 reveals that dark-skinned black women are given longer prison sentences than their lighter-skinned counterparts.

> The study, which sampled of over 12,000 black women imprisoned in North Carolina between 1995 and 2009, showed that light-skinned women are sentenced to 12 percent less time behind bars than their darker skinned counterparts. The results also showed that having light skin reduces the actual time served by 11 percent. (Smith 2011)

A University of Georgia study in 2006 revealed that employers prefer light-skinned black men to dark-skinned men, regardless of their qualifications (Banerji 2006). These results hint at the very fact that the majority of white mainstream Americans believe that blacks with lighter-skin have greater similarities than differences to white Americans as compared to their darker-skinned counterparts. According to Harrison, these similarities make white people feel more comfortable around light-skinned blacks, which translates into positive hiring practices for lighter-skinned blacks as compared to darker-skinned blacks (Smith 2011). As Harrison pointed out, skin tone differences are responsible for increasingly different perceptions within standard racially defined groups, like blacks (Banerji 2006). He highlights that other world cultures, such as Hindus in India, as well as some cultures of Asians and Hispanics are more aware of this color bias (Banerji 2006).

Many of these color biases and perceptions are also applicable to medicine, because doctors and researchers may tend to treat their lighter-skinned patients and research subjects with more respect and human dignity than their darker-skinned counterparts. As our society becomes more global, we must be cognizant of how our own biases affect decision making in areas such as medicine and health care. More sociomedical research needs to be conducted on the medical treatment and research selection of black patients and research subjects based on their skin color range (light to dark skin color) by doctors and researchers. These data results could help shed light on racial bias in the recruitment and selection of human subjects in clinical trials.

Recommendations and Conclusion

In my conclusion, I offer some relevant recommendations that may decrease the unethical practice of colorism in health care and medicine. First, as bioethicists, we must promote a genuine understanding about the origins of skin color and the reasons for skin color variation. The lack of understanding one's racial origins, in essence, leads one to make uninformed decisions and internalize false beliefs about one's own human value. The strategic promotion of self-esteem, self-efficacy, and positive body imagery can essentially lead to changes in one's psychological consciousness and behavior.

The arts can also serve as a means to help eliminate colorism in the study of human science. Earlier this semester, I visited the Metropolitan Museum of Art in New York City in order to conduct research on European artists' depictions of melanated people in Africa during the rise of European colonialism. In a marble and bronze sculpture entitled, *La Capresse Des Colonies* (1861), Charles Henri Joseph Cordier (1827–1905) portrayed his highly melanated female subject during his travels in France's North African colonies as exotic and beautiful, not in an exaggerated, stereotypical form as most European colonial artists did in their works.

The sculpture depicts the beauty of a tribeswoman who was a member of the highly melanated ethnic group of Africa and Madagascar, the Cafres. Cordier's admiration for the beauty of highly melanated people was unique and unusual during a period in world history when Europeans routinely portrayed Africans with dark skin as savages. To date, archeologists in northern Spain and throughout the Latin American world are still discovering monumental works by European, Moorish, and other craftsmen that highlight the strength, beauty, admiration, and grand genius of highly melanated people. As more of these accurate depictions of highly melanated people are revealed, negative perceptions of blackness will hopefully be reduced. Because of my own great admiration for Cordier's sculpture, I created a collage rendering based on his portrayal of *La Capresse Des Colonies* (see image 6). I believe that my own modern collage rendering of Cordier's sculpture will also help counter modern colorism.

Image 6
Collage Rendering: Obiora N. Anekwe

Colorism can be counteracted through psychosocial therapy in order to change the way in which one believes negative ideas about dark skin color. Through the analysis of the centralized root cause of colorism, there is a greater likelihood of holistic healing for all those affected by unjust discriminatory practices. Advocacy groups must also promote equality and the development of programs that counter colorism and promote the diversity of skin color (Lewis et al. 2013).

And lastly, there needs to be a conscious willingness to conduct research on skin color discrimination (colorism) in the medical treatment, practice, and clinical trial recruitment/selection of patients and research subjects by doctors and researchers. Such research will help uncover hidden race perceptions by medical professionals that can be resolved through continuing education. All in all, I believe that if racialized medical perceptions based on skin tone are reduced, perhaps the true intent of healing can be practiced in health care and medicine.

References

Banerji, S. (2006, August 31). "Study: Darker-Skinned Blacks Hit More Job Obstacles." *Diverse Issues in Higher Education*. Retrieved from http://diverseeducation.com/article/6306/.

Baruti, M.K.B. (2000). *Negroes and Other Essays*. Akoben House.

Brown-Glaude, W. (2007). "The Fact of Blackness? The Bleached Body in Contemporary Jamaica." *Small Axe* 24 no. 3, 34–51.

Bynum, E.B., A. C. Brown, R. D. King, and T. O. Moore, Eds. (2005). *Why Darkness Matters: The Power of Melanin in the Brain*. Chicago, Illinois: Images.

De Casanova, E.M. (2004). "No Ugly Women: Concepts of Race and Beauty among Adolescent Women in Ecuador." *Gender and Society* 18 no. 3, 287–308.

DuBois, W.E.B. (1903). *The Souls of Black Folk*. Chicago: A.C. McClurg and Company.

Eric, U., D. Nilanjana, E. Angelica, G. G. Anthony, and S. Jane. (2002). "Subgroup Prejudice Based on Skin Color among Hispanics in the United States and Latin America." *Social Cognition* 20 no. 3, 198.

Fanon, F. (1952). *Black Skin, White Masks*. New York: Grove Press.

Funderburg, L. (2013, October). "Changing Face of America." *National Geographic*, 82–84.

Glenn, E.N., Ed. (2009). *Shades of Difference: Why Skin Color Matters.* Stanford, California: Stanford University Press.

Herring, C. (2004). "Skin Deep: Race and Complexion in the 'Color Blind' Era." In Herring, C., V. M. Keith, and H. D. Horton. Eds. (2004). *Skin Deep: How Race and Complexion Matter in the "Color-Blind" Era.* Urbana: University of Illinois Press.

Jablonski, N.G. (2012). *Living Color: The Biological and Social Meaning of Skin Color.* Berkeley: University of California Press.

Khan, A. (2009). "Caucasian, Coolie, Black, or White? Color and Race in the Indo-Caribbean Diaspora." In Glenn, E. N., Ed. (2009). *Shades of Difference: Why Skin Color Matters.* Stanford, California: Stanford University Press.

Lewis, K.M., S. Harris, C. Camp, W. Kalala, W. Jones, K. L. Ellick, J. Huff, and S. Younge. (2013). "The Historical and Cultural Influences of Skin Bleaching in Tanzania." In Hall, R.E., Ed. (2013). *The Melanin Millennium: Skin Color as 21st Century International Discourse.* New York: Springer.

Luisa, N.B., I. K. Catarina, R. W. David, V. D. R. Ana, and G. L. Penny. (2006). "Self-Reported Health, Perceived Racial Discrimination, and Skin Color in s in the CARDIA study." *Social Science & Medicine* 63 no. 6, 1415.

Lynch, W. (1712). *Willie Lynch Letters: The Making of a Slave.*

Myrdal, G. (1944). *An American Dilemma.* New York: Harper & Row.

Painter, N.I. (2012, November 26). Lecture on, *The History of White People*. Harvard University Bookstore. Retrieved from YouTube.

Rogers, T. (2010). "The History of White People: What It Means to Be White." Salon Media Group. Retrieved from http://www.salon.com/2010/03/23/history_of_white_people_nell_irvin_painter/.

Smith, J.S. (2011, December 21). "Americans Rank Mixed Race People ahead of Blacks Socially." *The Gio*: An online magazine.

Stephens, D.P., and A. L. Few. (2007). "The Effects of Images of Women in Hip Hop on Early Adolescents' Attitudes toward Physical Attractiveness and Interpersonal Relationships." *Sex Roles* 56 no. 3-4, 251–264. doi: 10.1007/s11199-006-9145-5.

Thapar, R. (2008). *The Aryan: Recasting Constructs*. Gurgaon: Three Essays Collective.

Thompson, C. (2009). "Skin Tone and the Persistence of Biological Race in Egg Donation for Assisted Reproduction." In Glenn, E.N. Ed. (2009). *Shades of Difference: Why Skin Color Matters*. Stanford, California: Stanford University Press.

Trautmann, T. (2004). *Aryans and British India*. New Delhi: Yoda Press.

Ulen, E.N. (2013). *Dark-Skinned Vs. Light-Skinned: 500 Years of Self-Hatred*. DVD video.

Wade, T.J., and B. Sara. (2005). "The Differential Effect of Skin Color on Attractiveness, Personality Evaluations, and Perceived Life Success of s." *Journal of Black Psychology* 32 no. 3, 215.

Washington, H. A. (2007). *Medical Apartheid: The Dark History of Medical Experimentation on Black Americans from Colonial Times to the Present.* New York: DoubleDay.

Welsing, F.C. (2013b). Public Lecture On, "Surviving Racism in the 21st Century." YouTube Channel.

Wilder, J., and C. Cain. (2011). "Teaching and Learning Color Consciousness in Black Families: Exploring Family Processes and Women's Experiences with Colorism." *Journal of Family Issues* 32 no. 5, 577–604. doi: 10.1177/0192513x10390858.

The Global Phenomenon of Skin Bleaching: A Crisis in Public Health (Part I)
Obiora N. Anekwe, Ed.D, M.S (Bioethics)

The global issue of skin bleaching or lightening has become a public health crisis of epic proportions. Many people of color purchase and use skin-bleaching products that later cause skin discoloration, skin cancer, and other medical problems. Due to the depth of this phenomenon, I will devote a two-part series to this issue in order to explore the ethical and public health challenges involved in skin bleaching, especially involving people of color. The issue of skin bleaching is more dangerous and deadly for people of color throughout the world because the practice also affects psychological and physical faculties.

The psychological impact for many people of color of lightening their skin in order to fit within the larger society's definition of beauty is now more widespread than ever. Skin bleaching also symbolizes more complex psychological issues, such as self-perception and self-esteem that have plagued people of color since the advent of international slavery, especially in the Americas. For instance, African Americans are often coerced through the mass media to believe that lightening one's skin through bleaching

ISSUES OF RACE, JUSTICE, GLOBAL COLORISM, SKIN BLEACHING, AND INTERNALIZED PERCEPTIONS OF SKIN COLOR

brings greater acceptance in the larger North American society. But I believe coercion can be most effective unconsciously through culturally assimilative brainwashing, which may have a deeper impact on one's consciousness and decision to choose collective acceptance by the general society rather than individuality based upon positive views of oneself and one's own ethnic origins.

One of the most recent examples of coercive means to brainwash Africans throughout the West African diaspora is most visible in the phenomenon of bleaching one's skin in order for it to be lighter. This phenomenon has been described as medically risky and psychologically dangerous. Recent news reports throughout West and South Africa's media have reported on black women who buy skin-whitening products in order to bleach their skin tone to look whiter and, in their opinion, more beautiful. Although many African women may not view their actions as medically dangerous, numerous medical studies have already indicated that excessive usage of these skin-whitening products can have an adverse effect that may cause skin cancer. Some African nations have even considered limiting the sale of skin-whitening products because of the mass and, sometimes, excessive usage of these products by black women.

When interviewed by news outlets, many African women have expressed that media images portraying white women as beautiful greatly influenced their decision to whiten their own skin. Furthermore, these women have added that although skin-whitening products may have harmful effects, they would still use these products to look and feel more beautiful.[1] One has to wonder whether such a mass consumption of these products will eventually cause a self-induced public health crisis within a segment of the larger global population. The reality is that without the conscious awareness of the negative impact of some coercive measures and a strengthening of one's individual identity, those who are faced with similar issues will not be able to overcome similar challenges.

The ethical issue involved in skin bleaching is that many skin-bleaching products cause great harm to customers due to high levels of mercury. Thus, many customers who buy these products are endangering the health of their skin, while skin-care product companies continue to benefit financially from selling these products. In many instances, customers are uninformed about the medical risks involved in consuming excess amounts of skin-lightening creams, such as Artra. This product, in particular, is globally advertised to encourage women of color to purchase it in order to have lighter and more beautiful skin, according to popular standards of beauty.

Some critics point out that people of color who choose to purchase and use skin-bleaching products in excess are responsible for their own health risks. But as I see it, when a minority population consumes an excess amount of skin-lightening products, a significant global health care phenomenon is at risk. Physical harm and life-altering health problems may occur to people who otherwise would not face these challenges.

According to the World Health Organization (WHO), mercury is a common ingredient found in skin-lightening soaps and creams.[2] Although skin-bleaching products are hazardous to one's health and are banned in many countries, these products are available for sale over the Internet, providing unlimited access to potential customers. For example, in Mali, Nigeria, Senegal, South Africa, and Togo, 25, 77, 27, 35, and 59 percent of women, respectively, are reported to use skin-lightening products on a regular basis.[3] In 2004, nearly 40 percent of women surveyed in China (Province of Taiwan and Hong Kong Special Administrative Region), Malaysia, the Philippines, and the Republic of Korea reported using skin lighteners.[3]

ISSUES OF RACE, JUSTICE, GLOBAL COLORISM, SKIN BLEACHING, AND INTERNALIZED PERCEPTIONS OF SKIN COLOR

The most dangerous effect of the inorganic mercury contained in skin-lightening soaps and creams is kidney damage.[2] Mercury in these products may also cause physical reactions, such as scarring, reduction in the skin's resistance to bacterial and fungal infections, and also psychological outcomes such as anxiety, depression, or psychosis and peripheral neuropathy. [3-5]

More attention needs to be focused on restricting the sale of Internet skin-lightening products containing mercury because people of color, especially women, are being dramatically affected by exposure to these products. For instance, mercury in soaps, creams, and other cosmetic products is eventually discharged into waste water. The mercury then enters the environment, where it becomes methylated and enters the food chain as the highly toxic methylmercury in fish. Pregnant women who consume fish containing methylmercury transfer the mercury to their fetuses, which can later result in neurodevelopmental deficits in the children.[4]

Unfortunately, a generation of women is being negatively affected by excess usage of these products, and in many instances, the environment, including the water supply and food chain, is being dramatically compromised in countries where people of color live and consume skin-bleaching products. Therefore, it is reasonable to conclude that skin-lightening products have caused a silent but deadly epidemic in public health that needs to be addressed immediately. Although customers have the choice to purchase these products, the larger population is being affected healthwise. As bioethicists, it is our utmost responsibility to protect the larger population rather than to protect individual rights to purchase skin-bleaching products that contain mercury.

Ethical guidelines should be recommended by bioethicists to limit companies from selling skin-lightening products on the

Internet. These universal guidelines can be implemented by monitoring skin-care product websites and even restricting how many products a customer may purchase over a period of time. Such limits are only a first step in counteracting a systemic problem in public health. A collective effort should be initiated by bioethicists and public health officials globally to prevent companies from gaining high profits while harming the health of their customers. It will entail the international cooperation of health officials who can recommend ethical solutions to protect our citizens, especially vulnerable groups, such as women of color.

Education is one of the most effective ways to inform the general public about the hazards of skin-lightening products that contain mercury. Public service campaigns—through television, billboards, and the Internet—can inform customers about the risks of consuming these products and how they affect personal health, the general public, and our environment. Such awareness provides customers with the necessary public health information to make reasonable health care decisions.

In part two, in my next opinion editorial, I will briefly highlight how some international celebrities of color, such as Sammy Sosa, have been consumers of skin-bleaching products, only to negatively influence the general public's interest in buying these hazardous products. Their actions are also reflective of a larger psychological, physical, and global trend in attempting to lighten one's skin in order to gain acceptance by the ruling class. As a result, many people of color in nations such as Brazil and Nigeria have consciously and subconsciously equated beauty with whiteness. This trend is due, in large part, to the negative impact of European colonialism, which encouraged a divide-and-conquer mentality based on social classism, colorism, and racism.

Globally, women of color have also been increasingly consuming these skin-bleaching products that contain mercury because they unfortunately believe that their dark or brown skin is somehow unattractive or ugly. I also plan to argue that additional ethical and public health measures, based on psychological and educational dimensions, should be explored in order to decrease the usage of skin-lightening products containing mercury. In many instances, the general public is unaware of the role melanin plays in intelligence and soulfulness. Therefore, I will also explore how awareness of the power of melanin and race consciousness could help dispel the myth that whiteness or lighter skin equates to beauty and intelligence. In my opinion, such awareness could possibly reduce the public health phenomenon of skin bleaching.

Reference Notes

1. Nigerians in America (2013), "New Survey Says Nigerian Women Lead in Skin Bleaching," retrieved from http://cultureshocknigerians.com/news/new-survey-says-nigerian-women-lead-in-skinbleaching/#sthash.c9yf9i5Y.dpuf.

2. World Health Organization (2011), "Mercury in Skin Lightening Products."

3. UNEP (2008), "Mercury in Products and Wastes," Geneva, United Nations Environment Programme, Division of Technology, Industry and Economics, Chemicals Branch, http://www.unep.org/hazardoussubstances/LinkClick.aspx?fileticket=atOtPM-tTmU%3d&tabid=4022&language=en-US.

4. Glahder, C.M., P.W.U. Appel, and G. Asmund (1999), "Mercury in Soap in Tanzania," Copenhagen, Ministry of Environment and Energy, National Environmental Research Institute (NERI Technical Report No. 306), http://www2.dmu.dk/1_viden/2_publikationer/3_fagrapporter/rapporter/fr306.pdf.

5. Ladizinski B., N. Mistry, and R. V. Kundu RV (2011), "Widespread Use of Toxic Skin Lightening Compounds: Medical and Psychosocial Aspects," *Dermatologic Clinics* 29:111–123.

The Miseducation of Global Perceptions of the Negative Effects of Skin Bleaching (Part II)

Obiora N. Anekwe, Ed.D, M.S (Bioethics)

"I've been black and dark-skinned for many years. I wanted to see the other side. I wanted to see what it would be like to be white, and I'm happy" (Fihlani 2012). These words are those of South African musician Nomasonto "Mshoza" Minisi, who commented on her usage of skin-bleaching creams. Her sentiments are reflective of a shared consensus among many people around the world who share a universal standard of human beauty based on abstract, nonsubstantive values of selfhood and collective racial imagery. Unfortunately, a global misperception about dark skin color among some people still determines their self-esteem and confidence. I will discuss the damaging health care phenomenon of skin bleaching, describe how to help resolve this dangerous and unhealthy practice, and provide educational and ethical avenues by which solutions can be framed to help alleviate misconceptions about melanin.

In part one of my opinion editorials, I briefly discussed how Dominican baseball great Sammy Sosa succeeded immensely as a world-renowned athlete but still struggled with internalized issues related to his skin color. He ultimately lightened his skin in order to compensate for his dark-skinned complexion. His actions are reflective of a stigma that dark skin amounts to ugliness and minimal intelligence. Sosa should not be blamed for the increased usage of skin-bleaching products by consumers. He's simply a product of a highly coloristic society in the Dominican Republic that mostly believes that lighter skin brings forth greater opportunities. For instance, in a recent study of perceptions of skin color among Dominican children, most of them surveyed believed that whiter or lighter-skinned test dolls were more beautiful and intelligent than their darker-skinned counterparts (Cabrera 2014).

ISSUES OF RACE, JUSTICE, GLOBAL COLORISM, SKIN BLEACHING, AND INTERNALIZED PERCEPTIONS OF SKIN COLOR

Celebrities, such as Sosa, influence our social, economic, and health care behavior. As they lighten their skin with skin-bleaching creams, many unsuspecting customers also follow suit and buy similar products on the black market, often sold with higher levels of toxic mercury and other dangerous chemicals that harm the skin (Lewis 2007). One international celebrity, Cameroonian pop singer Dencia, lightened her skin and now sells her own line of skin-care bleaching creams called Whitenicious.

More than ever before, bioethicists must strive to inform potential buyers of the health risks of these products. They should also help educate dark-skinned people about the benefits of having higher levels of melanin, especially in tropical countries where melanin protects darker-skinned people from skin cancer (Welsing 2013a). Decades ago, Carter G. Woodson (1933) most eloquently concluded in his groundbreaking book, *The Mis-education of the Negro*, that dark-skinned people have been conditioned to believe that they are inferior simply based on the color of their skin. His words are still pertinent today. It is up to bioethicists to help reeducate highly melanated populations in order to effectively change negative misconceptions about dark skin color. In my estimation, this sort of reeducation may even change a persistent desire to buy skin-lightening creams that cause harm to the physical body. It is up to bioethicists to reframe these challenges and refocus on educating the public about how dark skin color and melanin benefit one's physical, mental, and spiritual essence.

Some opponents contend that teaching about the benefits of melanin is a form of pseudoscience and unrelated to the field of bioethics. To the contrary, I believe that teaching our most vulnerable about the significance of melanin is science in its purest form; for science is a search for truth free of deadly deceptions. I also argue that not to teach the most vulnerable about this pertinent matter would be ethically disingenuous since many such critics have used whiteness as the basis for genetic white superiority, even going as far as conducting race-based medical studies on vulnerable populations to prove their beliefs. Teaching about the benefits associated with higher levels of melanin builds human pride, confidence, and a positive consciousness for dark-skinned people who otherwise have been taught to believe that their dark skin color limits their destiny, value, and status as human beings. Most important, teaching highly melanated people about the medical benefits of dark skin may prove to be the most significant factor in reducing their need to bleach their skin.

But how can highly melanated people be reeducated about the negative health effects of skin-bleaching creams while reducing the sale of these products? Simply said, we need to first educate consumers. It begins with dispelling global preconceived myths about dark skin color through education and presenting a more balanced approach about the significance of melanin. For instance, many people are unaware of the origins of skin coloration and the creative functionality of the pineal gland. This organ is located in the central portion of the human brain. It produces melanin, the substance that generates skin color in the body (King 2001). Melanin also serves as a protective mechanism for people in tropical countries, such as Brazil, Jamaica, and the Dominican Republic (Welsing 2013b). When people of color in these regions use skin-lightening products, they abnormally reduce the ability of melanin to naturally protect their skin from diseases, such as skin cancer (Lewis 2007).

Although it is not widely discussed, skin pigmentation plays a vital role in one's human existence. Hence, the unnatural alteration of skin color through skin-lightening products causes greater harm than good to people of color who possess higher levels of melanin (Duke 2011). Located at the center of the brain, the pineal gland has also been referred to as the inner eye and the seat of the human soul, especially in Egyptian history (Nasheed 2012).

ISSUES OF RACE, JUSTICE, GLOBAL COLORISM, SKIN BLEACHING, AND INTERNALIZED PERCEPTIONS OF SKIN COLOR

According to some scholars, soulfulness is attributed to the pineal gland (2012). Melanin is, therefore, the neurochemical basis for what is called soul in black people (2012). In fact, the activation of the pineal gland is believed in biblical scripture to allow humans to meet God face-to-face (King 2001). In Genesis 32:30–31, this process is described in Jacob's transformative meeting with God: "And Jacob called the name of the place Pineal, for I have seen God face to face and my life is preserved. And as he passed over Pineal the sun rose upon him."

The pineal gland is also the first gland developed in the body (*The 8th Realm*, 2014). It is responsible for sensing light and darkness (inner of third eye); telling our bodies when to do its functions (biological clock); setting the pace or rhythm for the body (pacemaker); helping us orient ourselves (compass); and producing/secreting the hormones of serotonin and melatonin, which, in effect, creates melanin (2014). As a matter of fact, all humans have melanin but at different levels or classifications—six to be exact (2014). When people alter their melanin levels by unnaturally bleaching their skin, they deregulate the natural means by which the body regulates itself and preserves a healthy balance (Afrika 2009).

Education through public service announcements and the social media must also serve as a means to help eliminate the need to bleach one's skin. For example, Jamaica's public service campaign educates its citizens about the health risks of bleaching one's skin. For several decades, Jamaica has been plagued by a high number of instances in which people buy banned skin-lightening products. But within the past few years, public health officials in the country have decreased the black market sale of illegal skin-bleaching products through the "Don't Kill the Skin" prevention campaign (Lewis 2007). Even Dove, a leading manufacturer of soap and skin-care products, has released an internationally marketed Internet commercial on YouTube celebrating and redefining beauty by focusing on the strengths of color diversity among people of color. These campaigns seem to be effective in reaching a segment of the population who otherwise would be unaware of the dangers involved in skin bleaching.

Nations with high imports of banned skin-bleaching products should impose heavy fines for transporting and selling these products. Skin-care product companies should also begin to accurately portray the diversity and range of black and brown people throughout the world. The lack of diverse depictions in skin color only translates into inaccurate messages that beauty is found through having lighter skin color.

And lastly, bioethicists should dedicate more research to the investigation of melanin in order to debunk myths about the inferiority of highly melanated people. We must be willing to conduct research that challenges the nonscientific and racialized belief shared by some that highly melanated skin equates to human inferiority. This sort of belief system only increases the globalized consumption of skin-bleaching products. We must not only change the negative imagery of blackness, but we must also adopt a balanced perspective and knowledge of melanin's significance in relation to scientific inquiry and discovery.

Reference Notes

Afrika, L. (2009), *Melanin: What Makes Black People Black* (Long Island City, New York: Seaburn Publishing Group).

Cabrera, C.E. (2014, February 5), "Dominican Colorism," *Huffington Post: Latino Voices*, 1–6.

Duke, B. (2011), "Dark Girls: Real Women. Real Stories," Duke Media.

Fihlani, P. (2012), "Africa: Where Black Is Not Really Beautiful," BBC News: Johannesburg, South Africa, retrieved from http://www.bbc.com/news/world-africa-20444798.

King, R. (2001), *Melanin: A Key to Freedom* (Chicago: Lushena Books, Inc.).

Lewis, T. (2007), "Don't Kill the Skin Campaign Targets Illegal Bleaching Products," *Jamaica Observer*, 1–2, retrieved from http://www.jamaicaobserver.com/news/118126_Don-t-kill-the-skin-campaign-targets-illegal-bleaching-products.

Nasheed, T. (2012), "Hidden Colors 2: The Triumph of Melanin," *King Flex Entertainment: The 8th Realm: Online Journal of My Conscious Mind*; (2014) "The Blessing in Blackness: Thoughts on Melanin."

Welsing, F.C. (2013a), Public Lecture on "Can You Protect Your Melanin," YouTube Channel.

Welsing, F.C. (2013b), Public Lecture on "Surviving Racism in the 21st Century," YouTube Channel.

Woodson, C.G. (1933), *The Miseducation of the Negro*, Las Vegas, Nevada.

The Beauty of Lupita Nyong'o and Internalized Perceptions of Skin Color
Obiora N. Anekwe, Ed.D, M.S (Bioethics)

In an award acceptance speech earlier this year at the Seventh Annual Black Women in Hollywood Luncheon, Academy Award–winning actress Lupita Nyong'o, spoke about the global issue of skin bleaching within the black community. She won the award for her role as Patsey in the remarkable movie, *12 Years a Slave*. During the luncheon, Lupita read a letter from a black woman who stated that she wanted to use singer Dencia's Whitenicious cream to lighten her skin when Lupita, in her own words, "appeared on the world map and saved me" (*Essence Magazine*, 2014).

The woman Lupita spoke about represents a large number of highly melanated people who dislike their skin color and attempt to lighten it through skin-bleaching products. The award-winning actress also discussed how she prayed to God to erase her blackness and become light-skinned, but according to her, "He never listened" (2014). It was the internal self-assurance of Lupita's mother that convinced her that her blackness equated to beauty on the inside and outside. This reassurance, often associated with self-efficacy, can change the misguided perspective that dark skin is ugly and inferior scientifically, physically, and spiritually.

As Lupita's mother often reminded her, we should be cognizant that "you can't eat beauty. It doesn't feed you" (*Essence Magazine*, 2014). In other words, the consumption of skin-bleaching or lightening products cannot repair the internalized racism and racial stereotypes promulgated upon dark-skinned people. Lupita's mother is correct. Beauty cannot be consumed or acquired. It is the compassion for yourself and the world that surrounds us that defines us as humane. As Lupita portrayed Patsey in the movie, *12 Years a Slave*, she proved that even as the beauty of her body faded with time in the movie, the true essence and beauty of her spirit was still alive. And as Lupita put it so eloquently but plainly: "There is no shade in that beauty" (2014).

The story that Lupita highlighted speaks to the larger issue of colorism, which has plagued people of color psychologically, physically, and spiritually. The consumption of skin-bleaching products is an example of how negative perceptions about dark skin color externally manifest through detrimental physical actions. But more important, Lupita highlights that beauty is a redemptive quality that is measured by the eyes of its beholder. It is subjective, based on our own past experiences and preconceived notions of what constitutes *beauty*. As I have depicted in image 1, Lupita Nyong'o not only possesses physical beauty but a unique quality of internal gravitas that is rare among artists of her stature. This maybe the reason why the young lady Lupita spoke about may have deterred her wish to somehow become lighter skinned. She saw the internal resilience and humanity that Lupita possessed, which transcended her obvious external beauty. It is this quality that sustains us on the journey through life's many challenges. Ultimately, it is her internal self-confidence and efficacy that has brought her successfully to the world stage called *life*. We all can learn more than we realize from Lupita and the story she so vividly recounted about her newfound fan. We are a product of what we believe, but, more important, we are also the greatest determiner of our own destiny and our place within the world.

Reference Notes

Essence Magazine (2014, February 28), "Lupita Nyong'o Delivers Moving 'Black Women in Hollywood" acceptance speech, retrieved from http://www.essence.com/2014/02/27/lupita-nyongo-delivers-moving-black-women-hollywood-acceptance-speech/.

Image 1
Collage: Obiora N. Anekwe

We Are Our Brother's Keeper
Obiora N. Anekwe, Ed.D, M.S (Bioethics)

Young men of color, in particular, those of Hispanic and African American origins, have continued to constantly lag behind their peers in both education and health care. Although these two areas are of critical importance in achieving academic and life choice decisions, they have been ignored by society as integral components of domestic sustainability, access, and development. In many estimates, many young men of color are ignored and invisible to social welfare agents, such as teachers and health-care providers. Unconscious and even blatant forms of discriminatory practices in education and health care toward Hispanic and African American young men oftentimes negatively affects entire communities, mostly in urban and impoverished environments.

When these young men are not properly educated, they eventually are unable to build and reasonably sustain family structures because of a lack of the proper skills needed for long-term employment. These young men eventually become psychologically, spiritually, and physically frustrated and project their anger through criminal activities that harm their own communities. Rather than serving to advance society's moral and ethical good, they become an undeniable menace to society. In a capitalistic society such as ours, we are trained and even formally educated to believe that individualism is a paramount practice to success. But individualism can unfortunately also lead to greater detriments than good. Somehow or another, a purposeful lack of reasoned attention given to our most vulnerable youth has been manifested through violent crimes, imprisonment, mental depression, and health care problems (hypertension, prostate cancer, diabetes, and heart disease) that not only affect our young men later in life but our global community. When these physical, spiritual, and psychological abnormalities exist, educational and health-related advancement is less likely to occur.

So what is our social, moral, and ethical responsibility to provide proper education and health care to our most vulnerable population of young men? Our responsibilities must be multilayered, intentional, inclusive, and collective in scope and nature. We can first begin by implementing a number of progressive and even traditional initiatives that will sustain and promote the holistic development of our young men of color. These programs include mentoring programs, such as the Big Brother Mentoring Program; after-school programs; alternative teaching methods tailored to the learning needs of young men of color; preventive health and educational measures, such as mandatory free breakfast, lunch, and dinner meals in public and charter schools; mandatory prekindergarten education; coping strategy workshops and training programs; sexual reproductive health and mental/physical illness/disease prevention training programs; the advancement and establishment of male-gendered charter schools; psychological awakening and reeducation training initiatives; the reeducation of ancient African spiritualism; the training of core and traditional values, such as respect for persons and spiritual, physical, and mental cleanliness and purposeful living; and holistic healing initiatives.

These strategies and initiatives are a few examples of what can and should be done to help our young men. We have to come to a realization that what affects one segment of our society ultimately causes a ripple effect of hardship and burden to our entire global community. We are as strong as our weakest link. My recommended solutions to education and health care disparities among Hispanic and African American young men are both complex and challenging, but with persistence and ingenuity, these issues can be resolved. In the end, we all have to work together in order to save our youth from their greatest depths of despair.

Healing Voices of a People: Harry Belafonte and the Belafonte Folk Singers' Exquisite Interpretation of the Negro Spirituals and a Movement toward Modern Justice-Based Ethics
Obiora N. Anekwe, Ed.D, M.S (Bioethics)

The significance of the Negro spirituals was awakened in me from a recent visit to a local used record and bookstore in Brooklyn. During my exploration to purchase books on art and philosophy, I encountered an unanticipated surprise. Hidden within the shelves of records in the bookstore, I discovered a well-maintained record vinyl of Negro spirituals. As a youngster, I listened to my vinyls of Belafonte belting his native West

Indian calypso songs, such as "Day-O" ("Banana Boat Song") and "Jamaica Farewell."

It was not until I reached a later stage of maturity that I fully appreciated the healing voices of the Negro spirituals. As a youth, I was familiar with the transformative lyrics of spirituals from such oral truth-tellers as Mahalia Jackson and Harry Belafonte. But I gained a greater understanding of the universal messages translated in spirituals during my tenure as an undergraduate student and then as a graduate student at Clark Atlanta and Tuskegee Universities. These two institutions have historically been on the forefront of musical arrangement and composition in Negro spirituals from such musical geniuses as James Weldon Johnson (Atlanta University) and William L. Dawson (Tuskegee Institute).

Belafonte's immense stature as a vocalist is most evident in his monumental album on spirituals, *My Lord, What a Mornin'*, released in 1960 by RCA Victor. In this album, Belafonte moved into one of his most musically productive periods during the civil rights movement. The album primarily focused on specific folk-music themes. *My Lord, What A Mornin'* was the first of two albums featuring the choir known as the Belafonte Folk Singers, conducted by Bob Corman. The album consisted of traditional Negro spirituals, delivered by Belafonte, who combined his acting and singing abilities with his deep analytical understanding of West African moralism due mainly to his growing interest in his own Caribbean culture and the impending civil rights movement. Harlem Renaissance poet Langston Hughes wrote the album's liner notes, which described the Negro spirituals' historical origins.

In some songs, such as, "March Down to Jordan," listeners are metaphorically exposed to the lyrical messages among abolitionists and enslaved people that a journey down by the riverside would lead to freedom territories, such as Canada. Most slave owners only interpreted these lyrics as songs of temporary joy, but these songs intentionally possessed a much deeper meaning that an enslaved people would soon be free like the Hebrew people who escaped bondage in biblical Egypt. As Belafonte powerfully reiterates in song: "Have you heard of-a that city / They say it's built four square / He said He wanted you people / To meet Him over there." To know the lyrical content of such songs is to know our strength and genius as a people.

Black arrangers and composers of Negro spirituals interpreted a story of a people full of multilayered moral and ethical messages of social, medical, educational, economic, and spiritual freedom not yet possessed. It was hope that the spirituals provided in the mist of apparent despair. Freedom warriors and abolitionists, such as Frederick Douglass and Sojourner Truth, depended on the literal and figurative language of the Negro spirituals to interpret strategies for slave abolition and physical road maps for the Underground Railroad. Many white slave owners only heard the Negro voices singing the spirituals as abstract praises to God, but lyrics entrenched in these songs once uncovered hid literal coded messages of routes to freedom territories, such as Canada and Florida.

Dubbed the "King of Calypso," Harold George "Harry" Belafonte, Jr., was born in Harlem, New York, on March 1, 1927, to Melvine (née Love), a housekeeper of Jamaican descent, and Harold George Bellanfanti, Sr., a Martiniquan who worked as a chef in the National Guard. As it is told by Belafonte, he was working as a janitor's assistant in New York City during the 1940s when a tenant gave him, as a gratuity, two tickets to see the American Negro Theater. He fell in love with the art of acting and also met actor Sidney Poitier, who later became his lifelong friend.

As early contemporary actors in the American Negro Theater, Belafonte and Poitier became strategic financial supporters of the civil rights movement with their friend, Dr. Martin Luther King,

ISSUES OF RACE, JUSTICE, GLOBAL COLORISM, SKIN BLEACHING, AND INTERNALIZED PERCEPTIONS OF SKIN COLOR

Jr. In essence, these men quietly recruited other Hollywood celebrities to help sponsor the march on Washington in August 28, 1963. In the fight for civil rights during the 1960s, Belafonte employed the same methodical Underground Railroad tactics as his ancestors. Just as his ancestors sent freedom messages to enslaved blacks through Negro spirituals, so did Belafonte. He essentially served as the civil rights financial architect, smuggling money to civil rights workers throughout the South.

Today, Belafonte has dedicated the rest of his life to fighting injustice everywhere in every corner of the world in which it exists, from engaging in protests to end apartheid in South Africa to his efforts to help black and Latino gangs end violence in our most impoverished cities. He has chosen to fight these battles because it is the right thing to do. Often, he has criticized such celebrities as Beyoncé and Jay-Z for not being more vocal on issues that directly affect the black community. As Belafonte has noted, these celebrities have a moral obligation and ethical responsibility to fight for causes that matter, such as health care disparity and the disproportionate number of African American men and women who are diagnosed with prostate cancer and HIV/AIDS, respectively. They are ethically bound because of their positional power and their great influence on our emerging generation of African American and Latino youth. One reason why many popular musicians tend not to serve as advocates for the most vulnerable may be found in the messages they convey in their artistic works, which are often, more than necessary, shallow and negative.

During Belafonte's generation of entertainers, music seemed to be more aligned with and reflective of the struggles and critical issues of the day. The lyrical content of music was consistent with the struggle for freedom. In essence, music served as the great enabler to democratic progress. From the Negro spirituals to gospel to jazz to soul to rock and roll, these musical genres expressed the feelings of a people. They are also greatly rooted in the oral traditions of West Africa, often layered with moralistic and ethical meanings ingrained in cultural tradition and spiritual mythology. In America, the music of a people is rooted in the Negro spirituals.

Hundreds of years later, these songs continue to eloquently translate hidden textual meanings that freedom is always on the horizon and that it is always tangibly possible with faith and assurance. We must all strive to be as brave as our ancestors. They possessed a resilience beyond measure and expectation. In modern times, we still struggle for another form of liberty. Our enslaved ancestors fought for physical freedom, but today we struggle for mental freedom and the human right to acquire equalized educational resources and health care access. It is my greatest hope that we, as a people, renew our own spiritual essence through Negro spiritual interpretations in order to transform modern enslavement into transcendent freedom.

Chapter 2:

A TELEVISION SOAP OPERA'S CRITIQUE OF ISSUES RELATED TO HEALTH CARE INEQUALITY AND RISKY SEXUAL BEHAVIOR PRACTICES

Tyler Perry's *The Haves and the Have Nots*: Infusing Thematic Discussions of Health Care and Class Disparities through the Visual Medium of Television (Part I)

Obiora N. Anekwe, Ed.D, M.S (Bioethics)

Tyler Perry has become an American entertainment icon. His golden touch as an actor, writer, producer, and director has reached arenas in stage, television, and film. Perry continues to bring forth ethical and moral messages in the visual arts that are often overlooked or underestimated. It is no different with his first soap opera, *The Haves and the Have Nots*, broadcast on the OWN channel. The racially diverse cast of characters portrays the lives of the rich (the Cryer and Harrington families) and the poor and destitute (the Young family) in the southern city of Savannah, Georgia. But beyond the surface of class difference, these families share common bioethical themes and challenges such as cancer, substance abuse, homosexuality, suicide, prostitution, poverty, rape, perpetual crime, abortion, mental illness, and life support.

One of the more salient themes of the show is the issue of life support and health care disparities based on class. Katheryn Cryer (of the Haves) befriends and confides in Hanna, her maid (of the Have Nots), during her treatment for cancer. Katheryn and Hanna share a bond as parents, but most significantly, as two cancer survivors. After Hanna aids Katheryn in her recovery efforts, Katheryn promotes Hanna as the head maid in the Cryer household. Hanna has a son, Benny, who lives with her and diligently operates a tow truck company. While working late one night, Benny is accidentally struck by a car driven by Katheryn Cryer's son, Wyatt, who was under the influence of heroin. Wyatt not only harms Benny but subsequently kills a girl in the accident. As a result of this horrific turn of events, Benny is in a coma with very little hope of recovery. His mother, Hanna, constantly prays for her only son's recovery, while Wyatt, a young man of privilege, is being protected by his family's wealth and power. To make matters worse, Benny's father, Tony, who was not involved in his son's life as a young man, legally seeks to remove his son off life support in order to acquire a much-needed kidney.

This tale's tragedy is found in the victim of the accident, Benny. He cannot receive the treatment he deserves in a community hospital because he does not have health insurance. The lack of access to medical resources is limited by Benny's status as a young person without medical capital. Tyler Perry creatively brought the issue of health care disparities to the forefront without preaching to the choir. His writing is so well crafted that you, as a viewer, are confronted with this issue without much notice. Perry's narrative from *The Haves and the Have Nots* highlights how the passing of the Affordable Care Act begins to bring justice and equality to health care. Opposition to the act by those who seldom have read it in its entirety and oftentimes possess comprehensive health care insurance themselves simply brings to light subversive and undermining discriminatory practices against the poor and most vulnerable.

As Tyler Perry's television series demonstrates, the lack of comprehensive health insurance oftentimes results in the death of the poor. In this scenario, the poor and vulnerable are not the only victims. Their families and even the larger society are harmed because of a lack of compassion for the helpless. Rather than promoting further divide over disagreements about certain sections of the Affordable Care Act, we, as a nation, should seek to make health care a nationalistic fundamental right that benefits all society. Although the health care system in America seems fragile and even broken to some, it can be healed and repaired through a concerted effort of bipartisan political leadership and citizen support. Tyler Perry's television series reminds us that we have what it takes to bridge the divide between the Haves and the Have Nots if we are willing to work together for a common and necessary good.

Tyler Perry's *The Haves and the Have Nots*: Addressing Forced Duplicity in Black Male Identity and Risky Sexual Behavior Practices (Part II)

Obiora N. Anekwe, Ed.D, M.S (Bioethics)

Tyler Perry has done it yet again. As writer, executive producer, and director of the OWN channel's modern soap opera, *The Haves and the Have Nots*, Perry tactfully addresses the ethical challenge of forced duplicity in black male identity through the prism of risky sexual-behavior practices among African American homosexual men. For a number of years, many gay black men have been under societal pressure to conform and marry women. These practices eventually harm the entire family structure because many of these men live secret, closeted lives as homosexuals. In some cases, these men contract HIV/AIDS due to unsafe and risky sexual practices with other gay men, only to infect their wives or female partners. Within the black community, this practice is known as "living on the down low." Although this ethical issue is seldom discussed in the African American community, especially within the traditional black church, unsafe and risky sexual practices by married homosexual men still unfortunately affect the entire black family, especially black women, who make up one of the leading groups of HIV/AIDS victims.

In the season two episode "Again and Again," Tyler Perry highlights how "living on the down low" ultimately harms the medical good of society. The character, Jeffrey, wants to live free of judgment for his decision to tell his parents that he is gay. His father, David, is very supportive of his lifestyle choice. Jeffrey's mother, Veronica, on the other hand, is very judgmental and rejects his decision, only to forcibly blackmail him into living as a closeted homosexual young man. As a result, Jeffrey goes through a crisis of black male identity because he tries to appease his mother by living as a heterosexual. In order to make Jeffrey comply with her demands, his mother selects Melissa, a beautiful

African American young lady for him to date. Although Melissa is highly attractive, Jeffrey initially resists her sexual advances after their first dinner date only to have sexual intercourse with her later that evening. Although he does not desire her sexually, Jeffrey's risky behavior of having unprotected sex with Melissa contributes to the potentiality of acquiring HIV/AIDS.

Although the discussion of black male sexuality is a taboo subject within many black communities, it does not mean that the topic should not be discussed and confronted head-on. As long as this troubling trajectory persists, a growing number of African American women with HIV/AIDS will continue to negatively impact their children's upbringing. In the final analysis, children lose their primary caregivers because of this untimely disease.

Continuing educational health care initiatives through the black church is an effective means to help decrease the rate of African American women infected by HIV/AIDS because, in many cases, the makeup of this institution is mostly composed of black women. Although some black churches have initiated HIV/AIDS programs in recent years since the escalated rise of HIV/AIDS cases, many black churches are resistant to facilitating public discussions or program initiatives on this significant public health problem. A much more concerted effort by the black clergy must play a significant role in promoting health care awareness surrounding HIV/AIDS prevention and educational awareness. Because many facets of African American culture surround the black church, they are ethically bound and morally responsible for becoming change agents in the black community. The black church's deliberate influence in HIV/AIDS awareness and prevention can make the most positive impact in health care advocacy.

A false male identity based on stereotypical ideals of masculinity often poisons and deteriorates qualities of manhood that are otherwise valued. More salient values, such as care, discipline, empathy, and responsibility, are often replaced by a more abstract value system based on promiscuity, trickery, and thuggish mannerisms. Oftentimes, contemporary society and the mass media feed into these false notions of black male identity through nonsensical and stereotypical portrayals of black men in popular music and television. These portrayals only promote externalized black male toughness based on mentally enslaved beliefs of low self-worth and value. Tyler Perry's ingenious tackling of black male identity crafted in Jeffrey's storyline further shows that black men are complex and multilayered, full of contradictions like any other race of men. All in all, Perry's usage of HIV/AIDS awareness through a dramatic medium serves a multilayered purpose, bringing to light the health care dangers of "living on the down low" while being cognizant of the fact that this practice ultimately harms black men, the black community, and those they love the most.

CHAPTER 3:
HEALTH CARE AND MEDICAL ISSUES IN URBAN PUBLIC HEALTH

Rising Up: Hale Woodruff's Murals from Talladega College Travel the Country
Obiora N. Anekwe, Ed.D, M.S (Bioethics)

Hale Aspacio Woodruff (1900–1980) was a master of many arts, including printmaking, draftsmanship, and painting. I first heard of Woodruff as a first-year college student at Clark Atlanta University, where he taught art and founded the Atlanta University (now Clark Atlanta University) art department. One set of his murals titled *The Art of the Negro* is housed in the atrium of Trevor Arnett Hall on the campus. As both a student of public health and a visual artist, I was greatly influenced by his technique and skill as an artist and profoundly moved by his depiction of African American history.

Woodruff's legacy as a prolific art professor and visual historian is being remembered through the restoration of his murals from Talladega College through a nationally sponsored tour of *Rising Up: Hale Woodruff's Murals at Talladega College.*

I viewed his murals chronicling the history of Talladega College, the Amistad story, slavery, and other related themes at the 80WSE Galleries at New York University. To say seeing Woodruff's murals was an extraordinary moment would be an understatement. He not only traced the interpretive history of the Amistad saga in his large scale murals—brightly colored, historically sound, and distinctly original—but he portrayed how African Americans lived after slavery: in physical conditions that undermined their health and dignity.

Several of Woodruff's smaller pieces illustrate his concern with poverty, especially housing conditions of poor black families depicted in scenes of outdoor bathroom facilities, shacks, and community wells. As the artist chronicled—and many public health historians have confirmed—the urban population increased dramatically during the early 1930s. Fears within the white community of competition, decreased property values, and integrated schools further exacerbated the depredations of segregation.[1] These smaller works, such as "Results of Poor Housing," are just as artistically revealing and educationally moving as his larger murals in the traveling exhibition.

As Woodruff noted in the September 21, 1942, issue of *Time* magazine regarding his study of rural life in Georgia, "We are interested in expressing the South as a field, as a territory, its peculiar rundown landscapes, its social and economic problems, and Negro people."[2] Although Woodruff was one of the first academically trained African American art professors to document the public housing conditions of black people in the South during the Jim Crow era, *Time* magazine mistakenly coined his work in 1942 as "outhouse school of art."[3] But Ralph McGill, the antisegregation columnist and later editor of the *Atlanta Constitution*, wrote a more accurate portrayal of Woodruff's images of black urban life as art that speaks "out in rebuke. They are worth more, they say more than all the studies on economics and the need for slum clearance and for better housing."[4]

The exhibition of Woodruff's historic murals and smaller works will continue to travel the country until 2015. The next viewing of his art will be held May 16 through September 14, 2014, at the New Orleans Museum of Art. I encourage you to take time out and see this transformative exhibition of the murals and smaller works by artist Hale A. Woodruff.

Reference Notes

1. Hale Woodruff Bio, Accessed November 8, 2013, http://moyepto.com/programs/GSWA/artists/Woodruff,%20Hale/biography/Hale%20Woodruff%20Bio.pdf.

2. *Time*, September 21, 1942.

3. *Time*, September 21, 1942.

4. Ralph McGill, "Quiet Negro Artist Here Hailed as One of Modern Masters," *Atlanta Constitution*, December 18, 1935; quoted in Heydt, "Rising Up," 36.

Urban Ethics and Cultural Moralism Translated in Spike Lee's *Do the Right Thing*: Twenty-Five Years Later
Obiora N. Anekwe, Ed.D, M.S (Bioethics)

The year of 2014 commemorated the twenty-fifth anniversary of filmmaker Spike Lee's 1989 Brooklyn-based film, *Do the Right Thing*. As translated in the film's title, Lee challenges society to simply do what is right and morally good. Although he has often been criticized for portraying the Bed-Stuy (Bedford-Stuyvesant) neighborhood of Brooklyn, New York, through a romanticized lens, those who live outside my community seldom have the opportunity to see the expansive cultural diversity, transformative artistic intelligentsia, and richly textured historical legacy of Bedford-Stuyvesant. Typically, many people are only familiar with Bed-Stuy because it is similarly home to mega rappers Jay-Z (Shawn Corey Carter) and the Notorious B.I.G. (Christopher George Latore Wallace), including a host of other musical icons in hip-hop.

But recently, concern over the potential closing of the Interfaith Hospital in Brooklyn, New York, brought deserved attention to the urgent health care crisis in our community, especially among those who have often been medically disenfranchised. Human rights activist and legendary entertainer Harry Belafonte recently voiced his own concern about the possible closing of Interfaith Hospital, which brought much-needed attention to minority health challenges, such as prostate cancer, chronic diabetes, and heart disease.

Do the Right Thing presented and still relays a universal ethical message to find solutions to our most ingrained social challenges, such as white supremacy, poverty, health care disparities, and sexism through potent visual imagery and richly textured language. Lee's epic urban tale also powerfully highlighted how many urban minority households essentially became socially isolated and increasingly dependent on public welfare, which, in turn, destroyed the sustainable growth of black households, only to marginalize many black families into disparaging economic and psychological systems of modern Jim Crowism and entrenched enslavement. As the prison-industrial system in America has become the New Jim Crow, so have our health care and educational systems, where many vulnerable African American and Latino boys are isolated, ignored, and invisible, systematically heavily medicated, and methodically placed in special education classrooms until age appropriate to work as imprisoned laborers within the US penal system. Although health care and education are interchangeable human rights in most wealthy and highly industrialized nations, many Americans are forced daily to sacrifice their human dignity and rights to health care and education. These two ideals of a right to health care and education are grounded in and inextricably connected to human justice and fair allocation of resources. Although passage of the Affordable Care Act rightly sought to decrease health care inequality, it only serves as a beginning factor to full and expansive medical coverage for all American citizens.

In recent years, neighborhoods in Harlem, Bedford-Stuyvesant, Crown Heights, and Fort Greene have become more and more gentrified, bringing forth a newly transformed landscape of charter schools and better medical facilities for residents. But one has to wonder, why is it ethically and morally justified to ignore and sustain inadequate schools and hospitals before a large influx of mostly white residents when these issues have unfortunately existed for decades in our neighborhoods? Some twenty-five years later, *Do the Right Thing* keenly reminds us that we still have a long way to go before genuine human equality and dignity are radically valued.

www.ingramcontent.com/pod-product-compliance
Lightning Source LLC
Chambersburg PA
CBHW030754180526
45163CB00003B/1016